ESTIMATING COMBINED LOADS O
SOURCE POLLUTANTS INTO THE BORKENA RIVER, ETHIOPIA

Eskinder Zinabu Belachew

Thesis committee

Promotor
Prof. Dr K.A. Irvine
Professor of Aquatic Ecology and Water Quality Management
IHE Delft Institute for Water Education

Co-promotors
Dr P. Kelderman
Senior Lecturer in Environmental Chemistry
IHE Delft Institute for Water Education

Dr J. van der Kwast
Senior Lecturer in Ecohydrological Modelling
IHE Delft Institute for Water Education

Other members
Prof. Dr V.Geissen, Wageningen University & Research
Prof. Dr P. Seuntjens, Ghent University / VITO, Belgium
Dr B Bhattacharya, IHE Delft Institute for Water Education
Prof. Dr W.A.H. Thissen, TU Delft / IHE Delft Institute for Water Education

This research was conducted under the auspices of the SENSE Research School for Socio-Economic and Natural Sciences of the Environment

ESTIMATING COMBINED LOADS OF DIFFUSE AND POINT-SOURCE POLLUTANTS INTO THE BORKENA RIVER, ETHIOPIA

Thesis

submitted in fulfilment of the requirements of

the Academic Board of Wageningen University and

the Academic Board of the IHE Delft Institute for Water Education

for the degree of doctor

to be defended in public

on Tuesday, 26 March 2019 at 3 p.m.

in Delft, the Netherlands

by Eskinder Zinabu Belachew

Born in DebreBirhan, Ethiopia

CRC Press/Balkema is an imprint of the Taylor & Francis Group, an informa business

© 2019, Eskinder Zinabu Belachew

Published by:
CRC Press/Balkema
Schipholweg 107C, 2316 XC, Leiden, the Netherlands
Pub.NL@taylorandfrancis.com
www.crcpress.com – www.taylorandfrancis.com

ISBN ISBN: 978-0-367-25345-5 (T&F)
ISBN ISBN: 978-94-6343-561-1 (WUR)
DOI: https://doi.org/10.18174/466828

Acknowledgements

I would first like to thank those who have directly been involved in the process leading to this Thesis, my supervisory team, Professor Kenneth Irvine, Dr. Peter Kelderman and Dr. Johannes van der Kwast. They contributed tremendously, both as supervisors and co-authors. Their continuous dedication to the subject and encouragement have had a vital influence on my development as a researcher and the emerging of this Thesis, and offered me the opportunity to turn the data-poor and limited opportunity of studying typical sub-Saharan environmental issues in the industrializing catchment of Kombolcha city into this PhD Thesis.

Without the financial grant from the Netherlands Fellowship Programme (Nuffic), this Thesis would not have been realized, and I would like to give special thanks to the Netherland government. During my on-site studies in Kombolcha City, I received large field and data support from the Kombolcha Meteorological Directorate office, the Kombolcha Hydrology office and the Kombolcha City Administration office. The laboratory staff of IHE-Delft has been very helpful in the various laboratory analyses and procedures. I am especially grateful for the contribution of Mr. Fred Kruis, Mr. Ferdi Battes and Mr. Berend Lolkema. Perhaps the most direct assistance was given by those who helped me with the data acquisition in the field sites in Ethiopia. I want to extend my warm appreciation to Mr. Ali Seid, Mr. Beniam Getachew and Mr. Demissie Seid who were at times facing together the frightening lightening during rainstorms, snakes and thorns in the Kombolcha jungles and hyenas at night works. I enjoyed the help of MSc student Ahimed Seid from Ethiopia for collaborating in the analysis of sediment samples in IHE-Delft laboratory, The Netherlands. I also thank my employing organization, Wollo University, for their patience, encouragement and understanding to finish this Thesis.

Finally, apart from their encouragement, my family has always offered me the necessary peaceful working environment at home during these years. In particular, I would like to mention my dear wife Abeba Teklie. Without her good care for my son (Eyuel) and daughters (Maramiawit and Arsemawit), I would certainly have been too much strained.

Summary

Estimating the relative contribution of heavy metals and nutrients loads from diffuse and point sources and various hydrological pathways is a major research challenge in catchment hydrology. Understanding of the transfer, loads and concentration of these loads in basins is useful in designing and implementing policies for the managements of surface waters. In sub-Saharan countries, few studies have been done to estimate heavy metals and nutrients transfers in catchments. It is usually difficult to obtain hydrological and hydro-chemical data even for smaller catchments. This Thesis presents the estimates of loads of heavy metals and nutrients from industry and land use into two rivers that flow through an industrializing catchment, additionally presents the selection and application of a model to estimate TN and TP loads in the Kombolcha catchments. The study of the transfer contaminants from diffuse and point sources illustrates management, capacity and policy needs for the monitoring of rivers in Ethiopia, and with relevance for other sub-Saharan countries.

The study was done in the semi-arid catchments of Kombolcha city, which sits within an urban and peri-urban setting in north-central Ethiopia. The Leyole and Worka rivers drain the catchments, and receive industrial effluents from several factories in the surrounding area and wash-off from the surrounding catchment. The rivers flow into the larger Borkena River. The goal of this research was to monitor and quantify sources and transfer of heavy metals (Cr, Cu, Zn and Pb) and nutrients ((NH_4 +NH_3)–N, NO_3–N, TN, PO_4–P, TP) into the Leyole and Worka rivers, and evaluate their management/control in a data-poor catchment. The apportionments of the total nitrogen and phosphorus loads from diffuse and point sources were investigated. The work is placed in a policy context through a review of relevant policy within Ethiopia and at the wider perspective of sub-Saharan Africa.

The first set of measurements was on industrial effluent samples collected from discharge from five factories. In total, 40 effluent samples were taken in both 2013 and 2014. The second set of measurements were on waters and sediments. In total, 120 water samples were collected from the rivers in the wet season of two monitoring years of 2013 and 2014. River bed sediment samples, in total 18 samples, were taken at six stations on three occasions in the wet seasons the two monitoring years. In order to estimate the dilution capacities of the Leyole and Worka rivers, daily flow depths of the river water were recorded twice a day during the sampling campaigns of 2013 and 2014. Stage-discharge rating curves were used to estimate the flows of both the Leyole and Worka Rivers. The heavy metals concentrations were measured using Inductively Coupled Plasma Mass Spectrometry.

The median concentrations of Cr from tannery effluents and Zn from steel processing effluents were 26,600 and 155,750 µg/L, respectively, much exceeding emission guidelines. Concentrations of Cr were high in the Leyole river water (median: 2660 µg/L) and sediments (maximum: 740 mg/kg), Cu in the river water was highest at the midstream part of the Leyole river (median: 63µg/L), but a maximum content of 417 mg/kg was found in upstream sediments. Concentrations of Zn were highest in the upstream part of the Leyole river water (median 521µg/L) and sediments (maximum: 36,600 mg/kg). Pb concentration was low in both rivers, but, relatively higher content (maximum: 3,640 mg/kg) found in the sediments in the upstream of the Leyole river. Chromium showed similar patterns of enhanced concentrations in the downstream part of the Leyole River. Except for Pb, the concentrations of all heavy metals surpassed the guidelines for aquatic life, human water supply, and irrigation and livestock water supply. All of the heavy metals exceeded guidelines for sediment quality for aquatic organisms.

Regarding nutrients, emissions from a brewery and a meat processing unit were rich in nutrients, with median concentrations of TN of 21,00–44,000 µg/L and TP of 20,000 – 58,000 µg/L. These had an average apportionment of 10% and 13%, respectively, of the total nutrient loads. In the waters, higher TN concentrations were found from sub-catchments with largest agricultural land use, whereas highest TP was associated with sub-catchments with hilly landscapes and forest lands. Both the TN and TP concentrations exceeded standards for aquatic life protection, irrigation, and livestock water supply. Using specific criteria to assess model suitability resulted in the use of PLOAD. The model relies on estimates of nutrient loads from point sources such as industries and export coefficients of land use, calibrated using measured TN and TP loads from the catchments. The model was calibrated and its performance was increased, reducing the sum of errors by 89 % and 5 % for the TN and TP loads, respectively. The results were validated using independent field data.

The findings of the research shows high loads of heavy metals and nutrients in rivers of the industrializing regions of Kombolcha, identified gaps in estimating heavy metals and nutrient pollution and in policy implementation. Recommended future research and policy development to address a number of key gaps in water quality protection measures include control of point and diffuse loads of heavy metals and nutrients from sources, and improvement in land managements and monitoring and regulation.

Contents

List of abbreviations and acronyms

a.s.l	above sea level
BAT	Best Available Techniques
BMP	Best Management Practices
BOD	Biochemical Oxygen Demand
CCREM	Canadian Council of Ministers of the Environment
CEPG	Centre for Environmental Policy and Governance
COD	Chemical Oxygen Demand
Cr	Chromium
Cu	Copper
DAP	Di-Ammonium Phosphate
DO	Dissolved Oxygen
EC	Electrical Conductivity
EEPA	Ethiopian Environmental Protection Agency
EIDZC	Ethiopian Industrial Development Zone Corporation
EMEFCC	Ethiopian Ministry of Environment Forest and Climate Change
EMoWIE	Ethiopian Ministry of Water Irrigation and Energy
EMoWR	Ethiopian Ministry of Water Resources
EPLAU	Environmental Protection, Land Administration and Use
EQOs	Environmental quality objectives or standard
ESRI	Environmental Systems Research Institute
FAO	Food and Agriculture Organization
FDRE	Federal Democratic Republic of Ethiopia
FEPA	Federal Environmental Protection Agency
GRG	Generalized Reduced Gradient
GTP	Growth Transformation Plan
ha	hectare
ICP-MS	Inductively Coupled Plasm Mass Spectrometry
ILRI	International Livestock Research Institute
ISO	International Organization for Standardization
km	kilometre
mm	millimetre

MoFED	Ministry of Finance and Economic Development
N	Nitrogen
N.A.	Not Available
NE	North East
NH_3	Ammonia
NH_4	Ammonium
NO_3	Nitrate
OECD	Organisation for Economic Co-operation and Development
OM	Organic Matter
P	Phosphorus
Pb	Lead
PE	Polyethylene
PEC	Probable Effect Concentration
PLOAD	Pollution Load
PO_4	Phosphate
REPA	Regional Environmental Protection Authorities
S.C.	Sorting Coefficient
SDGs	Sustainable Development Goals
SIWI	Stockholm International Water Institute
SQG	Sediment Quality Guidelines
TDS	Total Dissolved Solids
TEC	Threshold Effect Concentration
TKN	Total Kjeldhal Nitrogen
TN	Total Nitrogen
TP	Total Phosphorus
TSS	Total Suspended Solids
UNEP	United Nations Environmental Program
USEPA	United States Environmental Protection Agency
USGS	United States Geological Survey
WHO	World Health Organization
ZID	Zone of initial dilution
Zn	Zinc

Chapter 1

1.1 General introduction

In many developing countries, water pollution is an ongoing and acute challenge for sustainable development. The transport of pollutants into surface waters has mainly increased because of anthropogenic factors (Hove et al., 2013; Alcamo et al., 2012; Crutzen and Steffen, 2003). As African countries gained political independence in the 1960s, they turned their attention to economic development mainly through industrial production and agricultural intensification (Steel and Evans, 1984). While many of these countries are committed to the 2030 Agenda for Sustainable Development and the Africa Union's Agenda 2063 (African Union, 2018), pressures to attract investors for industrialization and modern agriculturalization may reduce regard to progress with these Agendas (Xu et al., 2014; Sikder et al., 2013; Bertinelli et al., 2012). This Thesis focused on investigating the transfer of two groups of pollutants: heavy metals and nutrients in the rivers of a typical industrializing catchment of a sub-Saharan African city.

The term "heavy metals" refers to those metal and metalloid elements with relatively high densities ($>5,000$ kg/m^3). They are associated with eco-toxicity due to their non-degradable nature and accumulation in waters, sediments, and biota through the food chain (Goher et al., 2014; Xu et al., 2014). However, the term "heavy metal" is not always accepted; instead, some researchers recommend to just use the term "metal" (Duffus, 2002), (as used in the published paper of Chapter 2). This study focused on four heavy metals that include chromium, Cr (7,150 kg/m^3), copper, Cu (8,960 kg/m^3), zinc, Zn (7.134 kg/m^3) and lead, Pb (11,300 kg/m^3).

Nutrients includes the sum total of nitrogen (N) and phosphorus (P) that may be available in various forms. Total nitrogen may comprise nitrate (NO_3^-), nitrite (NO_2^-), ammonium ($NH_3 + NH_4^+$), and organic nitrogen (Kjeldahl-N). (N.B. charges will be left out in the various texts). Nitrite is generally unstable in surface water and contributes little to the total nitrogen. The main components of total phosphorus are soluble reactive phosphorus or orthophosphate ($PO_4^{3-} + HPO_4^{2-} + H_2PO_4^- + H_3PO_4$) and particulate phosphorus (PP). Dissolved phosphates are the most common forms of phosphorus found in in rivers where there are not large sediment loads. Phosphates are rather immobile in surface waters because of their strong attachments to soil particles. They can have a significant impact, however, because eroded soils can transport considerable amounts of attached phosphorus to surface waters. Too much N and P causes eutrophication and pollutes surface waters, with far-reaching impacts on public health, the environment and the economy (Delkash et al., 2018; EPA, 2017)

High releases of heavy metals and nutrients are a global challenge for surface water pollution (EPA, 2017; Landner and Reuther, 2004). The problems are increasing in sub-Saharan countries, arising from anthropogenic activities like industrial activity and intensive agriculture;   while monitoring and reporting on pollutant emissions are often absent, insufficiently reported, or of uncertain quality (Moges et al., 2016; Duncan, 2014; Mustapha and Aris, 2012). In Ethiopia, agriculture is the leading sector in the economy accounting for 43% of the country's gross domestic product. Increased food production through intensive agriculture is the primary goal of Ethiopian government policy (Awulachew et al., 2010). Also, government policy promotes a drive for industrialization, which stimulates growth of industries in specific zones throughout the country. Information on heavy metals and nutrients loads in rivers is often scant and their associated pollution risk unknown (Hove et al., 2013; Alcamo et al., 2012). The implication of the environmental policies is unclear and environmental institutions at regional and local levels are yet to be evaluated with respect to their roles for sustainable development (Sikder et al., 2013;Alcamo et al., 2012). With the fast industrialization and agricultural intensification and no understanding on effectiveness of regulatory structures and water quality monitoring, these problems will likely risk efforts towards environmental management and sustainable development.

1.2 Source and transfer of heavy metal and nutrient loads into surface waters

Heavy metal and nutrient loads can be released from diffuse (non-point) sources and point sources and transferred into surface waters (Novotny and Chesters, 1981). Quantifying the transfer and loads of these pollutants from their sources, and understanding the related managements are important, if environmental risks and hazards to receiving surface waters are to be addressed (Rudi et al., 2012). Characteristics of the receiving surface waters, like dilution capacity, pH and hardness of the receiving surface waters, influence the effects of the heavy metals and nutrients in the waters and are equally necessary to understand the associated risks (Pourkhabbaz et al., 2011; Ipeaiyeda and Onianwa, 2009; Besser et al., 2001).

In peri-urban environments, point sources usually comprise industrial effluent emissions and sewage treatment outflows. Depending on the raw materials and chemicals used in production processes, industrial effluents can contain, next to e.g. organic micro pollutants, heavy metals and nutrients. Ammonia, nitrate and phosphate are released by textile industries (Ghoreishi and Haghighi, 2003), while chromium, ammonia and organic nitrogen are released in tannery wastewater (Satyawali and Balakrishnan, 2008; Akan et al., 2007; Whitehead et al., 1997). Steel processing industries release effluents that are rich in metals (Rungnapa et al., 2010). The influence of these effluents to affect water quality depends on the extent of industrial activity and the level and the efficiency of pre-discharge treatment processes (Ometo et al., 2000). Although industrial pollutants entering to waters have been investigated worldwide (Landner and Reuther, 2004; Nriagu and Pacyna, 1988), they have yet to be assessed in many sub-Saharan countries (Oyewo and Don-Pedro, 2009). In these regions, in addition to the presence of relatively traditional and small-scale textiles and tanneries factories, there is a tendency to import cheaper technologies to cope with environmental requirements under increasing pressure of economical returns, often with treatment facilities that have low efficiency in reducing pollutants discharges to the waters (Rudi et al., 2012; Jining and Yi, 2009). This trend, which is realistically a "pollute now; clean-up later" action, may temporarily promote economic gains, but jeopardize the efforts to sustainable industrial development (Sikder et al., 2013; Alcamo et al., 2012; Rudi et al., 2012).

The diffuse sources of heavy metals and nutrients may comprise manures and commercial fertilizers in agricultural lands, weathering of rocks, and atmospheric deposition. The loads for these sources are transferred primarily during high rainfall events and enter into catchment streams with surface runoffs (Gil and Kim, 2012; Wang et al., 2006; Chiew and McMahon, 1999). The distribution of these pollutants into surface waters is affected by natural factors like precipitation, catchment surface characteristics (for e.g. topography and soil characteristics) and anthropogenic factors such as urbanization and land uses. The spatial variation of these factors affects their relationship with the hydrological chemistry of the streams in catchments (Johnson et al., 1997). The anthropogenic factors usually have greater impact on polluting surface water compared with natural processes (Hoos, 2008). However, both factors covary together, and hence, their combined effect has to be considered to understand the transfer of diffuse pollutants (Allan, 2004).

Land use intensification is a major anthropogenic factor that increases pollutants, especially nutrients, transfers into catchment streams (Gashaw et al., 2014; Griffith, 2002). The loading rate from each land use generally varies throughout the landscape depending on local factors such as precipitation, source activities, and soils (McFarland and Hauck, 2001; Johnson et al., 1997). Catchment-based water quality models mainly use such factors to estimate loads for management of water quality in catchments (Álvarez-Romero et al., 2014; Wang et al., 2013). In sub-Saharan countries, information on these factors are usually unavailable, even for the smaller catchments. While there can be temptation to invest in quite complex modelling, this does not necessarily result in a more accurate understanding of the underlying processes on which such models are based. The models can also be costly and subject to large errors in predictions from deficiencies in the data (Ongley and Booty, 1999). Therefore, starting with a basic model, for e.g. a generalized export coefficient of land uses (Soranno et al., 2015; Shrestha et al., 2008; Ierodiaconou et al., 2005), and gradually employing more detailed and comprehensive models, is a sensible approach.

With the presence of multiple point and diffuse sources into the pathway of surface waters, it is important to understand their loads and contribution, both as individual and combined sources. Many studies have examined industrial pollutants only from the perspective of the industry (Fuchs, 2002; Vink and Behrendt, 2002). In sub-Saharan countries, the attempt is customarily on reduction of point sources, neglecting other sources along pathways. However, the impact from a variety of sources can be significant and it is important to consider both the point and diffuse sources. Incorporating these sources is vital to include effects from the interaction among the complex system of water and landscapes and understand water flows through linked subcatchments in uplands and downstream lands that are far from the upstream lands. In this regard, catchment wide measurement of heavy metals and nutrients transfer into rivers is important to include sources and achieve a wider environmental benefit far beyond the obvious on-site and downstream impacts. With growing awareness of integrated catchment-scale natural resources in many African countries, (Darghouth et al., 2008), this has additional contribution to the advancement of global environment benefit.

1.3 The Borkena river basin and Kombolcha sub-basin in Ethiopia

1.3.1 Location, landforms, climate and land use

The study area of this research is located in the NE of Amhara Region, Ethiopia, between 11°4'59.74"N and 11° 4'44.14"N latitude, and 39°43'57.48"E and 39°39'31.26"E longitude (Figure 1.1.a, b). The Borkena river basin starts from the uplands of south Wollo Zone of the Amhara Regions and extends 300 km to the low lands of the Afar Region, draining an estimated area of 1735 km^2 (Figure 1.1.b). The basin comprises three hydrological sub-basins: upper (Dessie), middle (Kombolcha) and lower (Cheffa) sub-basins (Figure 1.1.c), and their main surface water drainage is controlled by the Borkena grabens that forms a regional linear drainage pattern. The Borkena River is the tributary of the Awash River, the largest river of Eastern Ethiopia (Figure 1.1.c.). The study area of this Thesis lies within 40 km^2 of the Kombolcha sub-basin including industrialized urban and peri-urban areas (Figure 1.1.c). The area is considered an ideal location for economic activities because of its intermediate location for domestic markets exports via the Djibouti port, which has been the only functional port to the land-locked Ethiopia (Figure 1.1.a).

The landform of the study area includes rolling and undulating hills, with high plateaus to the west, the Borkena graben in the centre and the southward sloping ground to the Borkena River (Figure 1.2.a). The elevation of the lands ranges from 1,750 m a.s.l. in the alluvial plain up to greater than 2,000 m a.s.l. in the uplands (Figure 1.2.b). Large parts of the built-up areas of the Kombolcha city have from 2.6 % to 10 % slope, and in the hilly areas, the slope increased to more than 20%. The local soils comprise alluvial/lacustrine deposits covering a large part of the town, with *Fluvisols* at the banks of the tributaries of Borkena, *Colluvial* screed deposits found mostly at the foot of hilly areas of the town and where *Cambisols* are developed, and Vertisol on the top of the Alluvial or *Colluvial* deposits, and covering most parts of the catchment areas (Zinabu, 2011). Several industrial effluents are discharged into the rivers of the Kombolcha catchment, eventually flowing into the Borkena River (Figure 1.2.a, b). The Leyole River receives effluents from industries including the steel processing factory, textile,

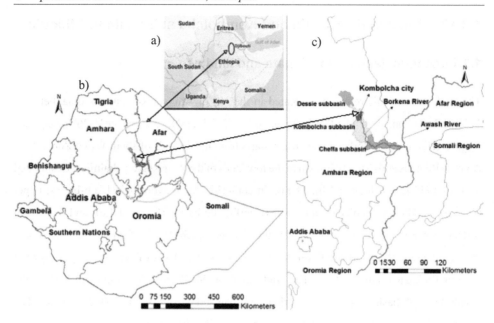

Figure 1.1. The map of the study area that is located in the horn of Africa, north-central Ethiopia (a), in the Amhara State (b), within the Kombolcha city administration, which is found in the Kombolcha sub-basin of the Borkena River basin (c)

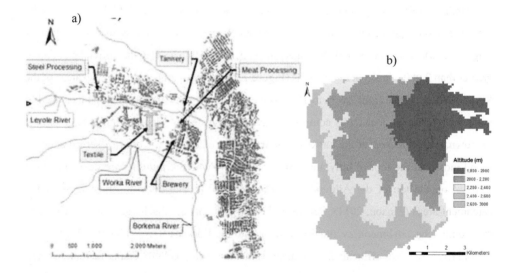

Figure 1.2. The location of the study area within the industrializing Kombolcha city administration including main rivers the Borkena River and its tributaries and factories discharging effluents into the Leyole and Work rivers (a), and surface land elevation of the Leyole and Worka rivers catchments in the Kombolcha sub-basin (b)

tannery and meat processing factory (Plate 1.1.b.), while a brewery discharges its effluent into the Worka River. The Kombolcha basin has a semi-arid climate. According to the Kombolcha Meteorological Branch Directorate report in 2013, the average annual rainfall is 1,030 mm, and the mean annual monthly temperature ranges from 24°C in January to 28°C in August. Kombolcha has two wet seasons, with the early wet season from February to April, and later in the summer from July to September. The rains in the early wet season have been very low in recent years because of recurrent droughts with high annual potential evapotranspiration, reaching up to 3,050 mm/year in 2014. (Kombolcha Meteorological Branch Directorate, 2015). The rainfall in the wet season of June to September has been remained relatively heavy and extensive (with a monthly average 710 mm) compared with the early wet season (having an average rainfall of 130 mm) (Kombolcha Meteorological Branch Directorate, 2015).

1.4 Problem statement and research framework

Based on the 2007 national census of the Central Statistical Agency of Ethiopia, Kombolcha district has a total population of 85,000. Industrial activities are notably one of the main economic forces in the urban, and agriculture is the main livelihood of the peri-urban and rural areas. Plantation forest and grazed land is common in the uplands of the catchments. Barren land is, however, evident in these uplands of catchments largely because of overgrazing and deforestation on the hillsides of the lands (Plate 1.1.c.) (Zinabu, 2011). The land use in the peri-urban area comprises crop and grazing lands, with moderate irrigation both up and downstream of the industrial areas (Plate 1.1.d.). The lower part (south-central) of the Kombolcha catchments consists of residential and industrial areas. Being topographically varied, both the rural upland landscape and lowland urban areas are prone to erosion (Plate 1.1.c). Diffuse loads are transported from these catchment areas into the Leyole and Worka rivers rising from the surrounding escarpments and draining eventually into Borkena River (Figure 1.2.a.). The hydrological flows of these rivers are modified by up-downstream agricultural irrigation and discharges of industrial effluents along the rivers (Figure 1.1.a.).

With abundant cheap labour force and opportunity for duty-free exports to the European and United States, many international investors are attracted to the city of Kombolcha and its medium to large-scale manufacturing industries. Currently, the Kombolcha is amongst the most industrialized areas of Ethiopia. Similar to many sub-Saharan cities, no study has yet been done to understand transfer of heavy metals and nutrients into the rivers of Kombolcha. This has contributed to often inadequate knowledge bases and local information for managements and protection of surface waters. Increasing such knowledge base is needed not only to overcome the information limitation but also to design and maintain environmental regulations.

Estimating the relative contribution of sources of the pollutants and the transfers of the pollutants into the rivers are important for planning and management of pollution loads (Bechtold et al., 2003). Licenced models that estimate loadings of heavy metals and nutrients are, however, often data demanding and costly. As is the case for many African cities, Kombolcha lacks access to these proprietary software and decision support systems due to limited finances (Rode et al., 2010; Loucks et al., 2005). Local authorities, therefore, lack the capacity to predict the loads of pollutants or measure their concentrations in rivers. Applicable models that are fit for data-poor situations and providing reliable information for the specific area conditions are needed to offset the burden of data dependency and costs.

Plate 1.1. Industrial effluent discharged via pipe-end into the Leyole River (a); industrial effluents transferred and spilling over the waters of the Leyole River (b); land use activities (e.g. Croplands, Grazing lands, Bare lands and Plantation forest) in the upper parts, with hilly landscapes, of the Leyole river catchment (c); and irrigation canal that diverts effluent mixed waters of the Worka river (d)

Like many African countries, Ethiopia has legislation to protect water resources for pollutions. The regulation to pollution of water resources follows a policy of "polluter pays" principle and is controlled by the Ministry of Water Irrigation and Energy and its sector institutions across regions. The industries in Kombolcha must follow the emission standards set in the *Environmental Pollution Control Policy (Proclamation No. 300/2002)* (EEPA, 2010). Also, pollution on water resources are regulated in accordance with the *Ethiopia Water Resources Management Regulation (No. 115/2005 and No.197/2000)* (EMoWR, 2004a). In Kombolcha, the risks of heavy metal and nutrient loads in the rivers remain unmeasured and the role for local environmental institutions in practicing the regulations is unexplained. With the rapid urbanization in the city, the impacts from new drastic changes of land use is not understood. Additionally, the environmental legislation is focused on controlling excess emissions from point sources and, like that for other regions in Ethiopia, the challenge with control of diffuse pollutants in the of Kombolcha catchment is yet to be fully addressed. Understanding the contribution of the diffuse loads in catchments is important to tackle the pollution problems and initiate developing relevant policies for the country. Effective pollution management of Kombolcha city, and those sharing similar situations, requires understanding of implication of the environmental policies and evaluation of the environmental institutions in regions with respect to their role for sustainable development.

As Ethiopia is signatory to the *Sustainable Development Goals (SDGs)* and has aligned the second *Growth and Transformation Plan (GTP-II)* to the goals (FDRE, 2016), the aim is to reach a full-fledged industrial development through expansion of food processing, garments and beverage industries that are mainly using raw materials from the country's vast agricultural production (MoFED, 2002). This means that Ethiopia is required to address detailed targets for pollution control and enhance regulation of pollution from sources. With the newly established industrial parks across the country and the government ambition to add more in near future, the industrializing city of Kombolcha can be a good test of the above commitments.

1.5 Research objectives

The objectives of this Thesis were to monitor and quantify the transfer of heavy metals (Cr, Cu, Zn and Pb) and nutrients ((NH_4+NH_3)–N, NO_3–N, TN, PO_4–P, TP) into the rivers and evaluate their management/control in an industrializing semi-arid catchments of the Kombolcha city in north-central Ethiopia. Specific research objectives were to:

1. quantify heavy metals (Cr, Cu, Zn and Pb) transfer, loads, and concentrations from industrial units into the rivers of the Leyole and Worka catchments, and assess regulations of industrial emissions into waters for Ethiopia and recommend related policy options;

2. quantify heavy metals transfer and concentrations into the water and sediments of the Leyole and Worka rivers, and review policies associated with water quality standards, compliance and recommend measures for improvement;

3. quantify nutrients (TN and TP) transfer, loads, concentrations and estimate apportionments of diffuse and point sources in the Leyole and Worka rivers catchments;

4. screen a number of water quality models that adequately estimate annual total nitrogen and phosphorus loads for the data-poor Kombolcha's catchments and simulate the changes in the loads due to *Best Management Practices* (*BMP*); and,

5. recommend monitoring and management of heavy metal and nutrient transfers into rivers and improvement for policy options and future studies.

These objectives supplement filling the gaps in understanding that are mentioned in *section 1.4*. More specifically, they will contribute to improvement of applicable methods to quantify loads of diffuse and point sources in data-poor areas, increase knowledge about impacts of industrial and agricultural land uses, and identification of gaps in controlling emission changes and providing policy options for improvement in rivers water protection.

1.6 Thesis structure

This Thesis consists of six chapters. **Chapter 1** presents the general introduction, highlighting current gaps in Ethiopia in understanding heavy metals and nutrients loads into surface waters. This is followed by presenting the problem statement, regarding heavy metals and nutrient loads in the Kombolcha catchments, and research framework and objectives of the study. A description of the study area includes information about the location, landforms, soils, and climatic data of the Kombolcha catchments, Ethiopia. **Chapter 2** explains the first research objective by quantifying the heavy metals concentrations and loadings from industrial effluents that are discharged into the Leyole and Worka rivers, and evaluates the industries compliance with water quality guidelines. This chapter also aims at contributing to the second objective of the research, by identifying gaps in industrial pollution control and recommending policy options. **Chapter 3** contributes to achieve the third objective of the research. This chapter quantifies the heavy metals transfer and concentrations in the Leyole and Worka rivers water and sediments, and illustrates review of application for river water quality standards and compliance. Recommendations for improvement are included in the chapter to fulfil the third research objective. **Chapter 4** is devoted to the fourth objective and quantifies the transfer and concentrations of nutrients into the Leyole and Worka rivers. This Chapter describes source apportionment of nutrients loads within the rivers catchments, influence of land cover and human open defecation and point sources on the transfer of nutrients, and identifies gaps in nutrient pollution assessment in rivers in order to use information for reconciling land use intensification with development goals. **Chapter 5** is dedicated to the fifth objective of the Thesis. The chapter screens a number of models that can estimate the annual TN and TP loads and are applicable to the semiarid and data-poor Kombolcha catchments. The Chapter also estimates the changes in the TN and TP loads due to best management practices using a calibrated applicable model for the Kombolcha catchments. Finally, **Chapter 6** (Synthesis and Conclusions) discusses and integrates the results of the previous Chapters. Key factors affecting the heavy metals and nutrient transfer in the Leyole and Worka rivers catchments are explained and management and policy options for improvement and future studies are proposed.

Chapter 2

Impacts and policy implications of heavy metals effluent discharges into rivers within industrial Zones: A sub-Saharan perspective from Ethiopia

Publication based on this chapter:

Zinabu E, Kelderman P, van der Kwast J, Irvine K. 2018. Impacts and policy implications of heavy metals effluent discharges into rivers within industrial Zones: A sub-Saharan perspective from Ethiopia. Environmental Management, 61:700-715.

Abstract

Kombolcha, a city in Ethiopia, exemplifies the challenges and problems of the sub-Saharan countries where industrialization is growing fast but monitoring resources are poor and information on pollution unknown. This study monitored heavy metals Cr, Cu, Zn, and Pb concentrations in five factories' effluents, and in the effluent mixing zones of two rivers receiving discharges during the rainy seasons of 2013 and 2014. The results indicate that median concentrations of Cr in the tannery effluents and Zn in the steel processing effluents were as high as 26,600 and 155,750 µg/L, respectively, much exceeding both the USEPA and Ethiopian emission guidelines. Cu concentrations were low in all effluents. Pb concentrations were high in the tannery effluent, but did not exceed emission guide-lines. As expected, no metal emission guidelines were exceeded for the brewery, textile and meat processing effluents. Median Cr and Zn concentrations in the Leyole river in the effluent mixing zones downstream of the tannery and steel processing plant increased by factors of 52 (2660 compared with 51 µg Cr/L) and 5 (520 compared with 110 µg Zn/L), respectively, compared with stations further upstream. This poses substantial ecological risks downstream. Comparison with emission guidelines indicates poor environmental management by industries and regulating institutions. Despite appropriate legislation, no clear measures have yet been taken to control industrial discharges, with apparent mismatch between environmental enforcement and investment policies. Effluent management, treatment technologies and operational capacity of environmental institutions were identified as key improvement areas to adopt progressive sustainable development.

2.1 Introduction

In many sub-Saharan countries, water pollution is an ongoing and acute challenge for sustainable development (Hove et al., 2013; Alcamo et al., 2012). Environmental regulatory structures may be in place, but pressures to attract investors for industrial activities may reduce regard for pollution abatement (Xu et al., 2014; Sikder et al., 2013; Bertinelli et al., 2012). Policies to promote economic gains can lead to a path of "pollute now; clean-up later" (Sikder et al., 2013; Alcamo et al., 2012; Rudi et al., 2012). The seemingly existing paradox of crafting good environmental policies but low enforcement has a risk of making the industrial growth unsustainable. Also, many industrial technologies are quite old and there is a tendency to import cheaper technologies to cope with environmental requirements under increasing pressure of economical returns (Rudi et al., 2012; Bertinelli et al., 2006). According to the environmental Kuznets curve (Grossman and Krueger, 1991), the ratio of socio-economic development to pollution may increase till the technology reaches the scrapping age, when operational cost can no longer cover market value for environmental quality (Bertinelli et al., 2012). Thereafter, this ratio will decrease only if improvement of the technologies reduce environmental impact. Industrial effluents containing heavy metals, and their accumulation in sediments and biota, present a persistent threat to ecosystems health (Xu et al., 2014; Kelderman, 2012; Jining and Yi, 2009; Gaur et al., 2005). This holds also for sub-Saharan African countries, where regular monitoring is limited (Ndimele et al., 2017; Akele et al., 2016). Thus, identifying effluent concentrations and discharge management are of increasing importance if environmental risks and hazards are to be addressed (Rudi et al., 2012).

This study is focused on the industrial city of Kombolcha in Ethiopia (Figure 2.1.), a typical sub-Saharan African city where urbanization and industrialization are growing fast but monitoring resources are poor. Data for industrial effluents and water quality are scant here and the threat to sustainable development is unknown. Backed by the government, the city's industries are growing fast, with the expansion of existing ones and ambitions to attract foreign investors for new ones. These industries discharge effluents into nearby Waterways. While industrial pollution control policies have been formulated for the country, the environmental institutions at regional and local levels are yet to be evaluated with respect to their role for sustainable industrial development. In this study, we examined the dissolved heavy metals: chromium (Cr), zinc (Zn), copper (Cu) and lead (Pb) in the effluents of five industries. The study aimed to (1) quantify the metal concentrations and loadings from these industrial effluents; (2) assess metal concentrations in the effluent mixing zones of the receiving rivers; and (3) evaluate the industries compliance with water quality guidelines, and identifying gaps in pollution control to recommend policy options.

2.2 Materials and Methods

2.2.1 Study Area

Kombolcha, in the North central part of Ethiopia, covers 125 km^2 (Figure 2.1.a), comprising rural upland landscapes in the north and populated lowlands in the south. Different land use types exist in the area, with extensive agriculture and forest land in the upland zone, and peri-

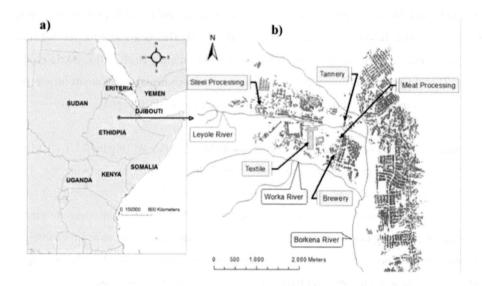

Figure 2.1. Location of the study area: a) Study area in East Africa, northern Ethiopia, b) Kombolcha industrial area (source: Kombolcha Administration City Office (2014))

urban and heavily urbanized and industrial areas mainly in lowland plains. The soils of the study area are generally Vertisol while the river banks and the foot of upstream hills are dominated by *Fluvisols* and *Cambisol* soil types, respectively (Zinabu, 2011). The area has annual bimodal rainfall seasons, usually from February to April, with heavier rainfall from July to September. Several tributary rivers rise from the surrounding escarpments and drain into two rivers, the Leyole and Worka rivers, which flow through an industrial zone of Kombolcha (Figure 2.1.b). The Leyole River receives effluents from the following four factories (Figure 2.2.):

- Steel processing factory, producing 26,000 tons per year of corrugated iron sheet;

- Textile factory, producing 22 million textiles per year, in garment production and dyeing;

- Tannery (not operating in 2013), soaking 1000 sheep skins and 3200 goat skins per day;

- Meat processing factory, dressing maximally 200 cattle per day.

The Worka River receives effluents from a brewery factory, with a production capacity of 250,000 bottles of beer per day (330 mL beer per bottle).

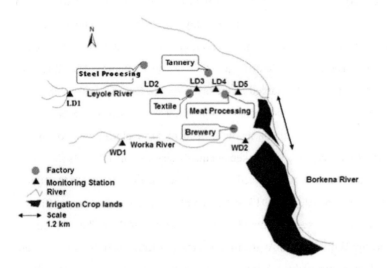

Figure 2.2. Schematic outlines of the rivers receiving the effluents of five industries, the factories' effluent discharge points and the monitoring stations and codes (LD1 (Confluence point of upper part tributaries and start of upstream Leyole river); LD2 (Steel processing effluent mixing zone in the Leyole river); LD3 (Textile effluent mixing zone in the Leyole river); LD4 (Tannery effluent mixing zone in the Leyole river); LD5 (Meat processing effluent mixing zone in the Leyole river); WD1 (Upstream Worka river); and WD2 (Brewery effluent mixing zone in the Worka river) along the Leyole and Worka rivers flowing into the Borkena river

2.2.2 Sample Collection, preservation and analysis

Factories effluent and river water sampling

Sampling was done in two bimonthly (15/30) monitoring campaigns during the rainy season from June–September in 2013 (campaign C1) and 2014 (C2). Samples were taken to measure total dissolved Cr, Cu, Zn, and Pb directly in the five factories effluents. Additional monitoring took place in the effluents mixing zones of the Leyole and Worka rivers (LD2-5; WD2; Figure 2.2.). Stations at the confluence of three tributaries in the upper part of the Leyole river (LD1) and confluence of two tributaries in the upper part of the Worka river (WD1) were located upstream of the industrial zone. The latter provides a theoretical baseline for estimates of pollution from industrial effluents. Results were used in evaluating river water quality.

pH and EC (electrical conductivity) were measured in situ using a portable pH (WTW, pH340i) and EC (WTW, cond330i) meter, respectively. The industrial effluent samples were taken directly from the discharge pipes using a 100 mL polyethylene (PE) sample container. Grab samples were taken based on equally spaced time intervals or volume, and were then mixed to make a composite sample. The choice for either equal time or equal volume sub-samples was based on the way the effluents were discharged by the factories. For factories with intermittent batch discharge of process effluents, three grab samples were collected at the beginning, halfway through, and at the end of the discharge of the effluent. For factories with continuous effluent discharges, eight grab samples were taken at equally spaced time intervals (i.e. every 3 hr in a 24 hr period), and samples were mixed in equal batches. In total, 40 (8*5) effluent samples have been taken in both 2013 and 2014. Water samples were also taken in the effluents mixing zones within a 5 m long section immediately downstream of the effluent discharge points into the Leyole and Worka rivers. As the mixing zones of Kombolcha's factories were not exactly determined, we assumed that a 5 m long section was sufficient for complete mixing of the effluent, containing both the zone of initial dilution (ZID), near the effluent outfall, and the chronic mixing zone (impact zone) (Alonso et al., 2016; Schnurbusch, 2000). The samples were collected at 1/4, 1/2, and 3/4 of the width of the river, and a composite sample prepared from equal volume proportions in a 100 mL PE container. Thus eight river water samples were taken for the two monitoring campaigns at the seven stations (Figure 2.2.), yielding a total of 56 samples in both 2013 and 2014. Both the river water and effluent samples were preserved with 1 mL concentrated H_2SO_4 to keep the pH < 3 in order to prevent metal adsorption onto the PE container wall (Rice et al., 2012). Within 15 to 135 days, samples were air-transported to the IHE Delft laboratory, located in Delft the Netherlands and kept in a cold room (<4 °C). Following the ISO 5643-3 guideline, the samples preservation time was always <6 months (ISO, 2003). A 10 mL sub-sample of the effluent was then filtered over a Whatman GF-C glass microfiber filter (pore size 1.25 µm) and diluted with Milli-Q water. The heavy metals concentrations were measured using ICP-MS (Inductively Coupled Plasma Mass Spectrometry), XSERIES 2 IUS-MS. All analyses were done in accordance with APHA-AWWA-WPCF-2012 (Rice et al., 2012).

Hydrology Measurements

The industrial effluent discharges were measured while collecting the above water samples, using the volumetric method, a simple and accurate method for very small flows with free-fall, such as at the outfall of a pipe or culvert (Hamilton, 2008). The time to fill a known volume (40 L) of effluent container was first estimated for each factory's effluent discharge pipe and flow rates were calculated by dividing the volume by the time to fill the container.

In order to estimate the dilution capacities of the Leyole and Worka rivers, daily flow depths of the river water were recorded twice a day during the sampling campaigns for four months, from 1 July to 30 September 2013 and 2014, in line with Herschy (1985). The measurements were taken at LD1, LD5 and WD2 (Figure 2.2.). In addition, 12 discharges were measured in three flow regimes (low, medium and high flows) following the methods outlined in ISO regulation 1100-2 (Voien, 1998). The river channel cross-section was first divided into vertical subsections. In each subsection, the area was estimated by measuring the width and depth of the subsection, and the water velocity was then determined using a current meter (Price-Type AA) or a pigmy-current meter. For low flows and shallow water depths at the start (i.e. in June) of the campaigns, a pigmy meter was used, whereas a vertical axis cup current meter was used for medium to high flows. The discharge (m^3/sec) in each subsection was computed by multiplying the subsection area by the measured velocity, and the total discharge estimated by summing up the discharges for each subsection. Stage-discharge rating curves were then prepared following Kennedy (1984). The least mean square method was used to estimate rating curve coefficients and, from that, the flow rates of the Leyole and Worka rivers (Das, 2014).

Statistical Techniques

All water quality data analyses were performed in R statistical packages (R Core Team, 2015). Normality of the data was first tested using a Shapiro-Wilk normality test (Degens and Donohue, 2002; Shapiro and Wilk, 1965), in order to choose the required statistical methods for further data analysis. Descriptive statistics were carried out for the results of the sample analyses. Here the data set for each station was found to be asymmetrically distributed with the mean values affected by a few high or low values (Table 2.1.). To best summarize these data sets, median values were selected for better representation of central tendency concentrations at each station (Bartley et al., 2012). These median values were compared with environmental guidelines.

Metals Mass Transport Loadings

The loading (g/day) estimations were computed in a Flux 32 software environment, an interactive computer programme used to estimate the loadings of water quality constituents such as nutrients, heavy metals and suspended sediments. The software incorporates six methods of estimating loadings of water quality constituents (Walker, 1990; Walker, 1987). As loadings by the factory effluents are not expected to vary much with effluents flows, a "direct loading median" method was used by determining medians of the loadings of a metal at each sampling time. These were derived from median of the product of metal concentrations and effluent of the factories during each sampling. The method is somewhat different from "numeric integration" which is based on the average of the loadings at each sampling time (Walker, 1987). Similarly, the loadings in the effluent mixing zones of the rivers were estimated using the product of median concentrations of the heavy metals and average flows of the river at a station. This method is appropriate for cases in which concentrations of heavy metals tend to be inversely related to flows, and loadings do not vary with river flow (Walker, 1987). This often occurs at effluent mixing zones for industries, as the flow and concentration relationships are controlled by dilution (Walker, 1990; Walker, 1987).

Quality Assurance

Quantification of heavy metals concentrations was based on calibration curves of standard solutions of the heavy metals. Detection limits were: Cr: <0.07 µg/L; Cu: <0.01 µg/L; Zn: <0.1 µg/L; Pb: <0.06 µg/L. The precision of the analytical procedures expressed as the relative standard deviation (RSD) was 5–10%. The ICP-MS measurements always had an RSD of <5%. For all samplings, blanks were run and corrections applied, if necessary. All analyses were done in triplicate.

2.3 Results

2.3.1 Discharges of the Leyole and Worka rivers

The hydrological flows of the Leyole and Worka rivers are modified by midstream industrial effluents and up-downstream agricultural activities along the rivers (Figure 2.1.). Though the rivers are having a width >4 m and depth of 3–5 m, the flowing water depth and width were not more than 1.25 and 2 m, respectively. For both 2013 and 2014 (Figure 2.3.), highest discharges were observed at all stations in July and August, as a result of increased rainfall, and reaching maximum discharge rates of approximately 0.9 and 1.3 m^3/s in the Leyole and Worka rivers, respectively (Figure 2.3.).

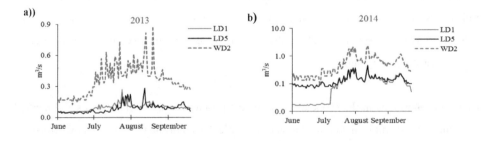

Figure 2.3. Water flows of the rivers. a Water flows (m³/s) of upstream Leyole River at station LD1, b downstream at station LD5, and c at the downstream Worka River station WD2, from 1 June to 30 September 2013 and 2014. Note the logarithmic scale in Figure 2.3.b

In the upstream part of the Leyole river (just downstream LD1; Figure 2.2.), daily mean flow rates ± standard error (n = 122) in the rainy seasons of 2013 and 2014 amounted to 0.12 ± 0.01 and 0.18 ± 0.11 m^3/s, respectively. For the downstream part of the Leyole river, at LD5 (Figure 2.2.), these values were 0.14 ± 0.02 m^3/s. and 0.28 ± 0.28 m^3/s, respectively. Comparing the upstream and downstream flows, the dilution factors for the average flows in the downstream zone of the Leyole River amounted to 45% in the 2013 and 61% in the 2014 campaign. Similarly, based on campaign comparison, the dilution factor increased in 2014 by 88 to 108% upstream and downstream for the Leyole River, respectively. For downstream Worka river, at WD2 (Figure 2.2.), the mean daily flow rates were 0.36 0.05 and 1.3 ± 0.1 m^3/s, for the rainy season of 2013 and 2014, respectively (Figure 2.3.). The low river discharges reflect the area's semi-arid climate.

2.3. 2 Metals in the Effluents and Effluents Mixing Zones of the River Waters

Heavy metals in the effluents of the five factories

In the following, the 2013 campaign will be indicated as C1, the 2014 campaign as C2. In virtually all cases, the metal concentrations were distributed asymmetrically, with mean values affected by a few high or low values (Table 2.1.). The EC for the steel processing effluent was found to be higher than for the other factories effluents (Table 2.1.), though the high values of the standard errors make it hard to give definite conclusions. This effluent was also acidic, probably because of pickling acids used to remove oxides from steel surfaces. In contrast, the effluent from the brewery was alkaline, attaining pH values > 11, likely coming from detergents used for washing equipment.

Metal concentrations in the effluents were often characterized by high extremes and marked differences between mean and median values (Table 2.1.). For the C2 campaign, Cr (median: 26,800 µg/L; maximum: ca. 65,000 µg/L) was very high in the tannery effluents compared with the other factories' effluents. The relatively low Cr (median: 6.1 µg/L) contents in tannery effluents for campaign C1, and corresponding Cr loadings (Table 2.1.), can be ascribed entirely to the cessation of the tanning processing during this first campaign. Cr median concentrations in the tannery factory effluents (Table 2.1.) exceeded the guidelines values of both USEPA and EMoI. In contrast to Cr, Cu effluent concentrations were below the two quality guideline values for all factories, but with noticeably higher concentrations in the steel processing factory than in the other effluents. However, owing to larger effluent water discharges, higher Cu effluent loadings (g/day) were periodically observed for the tannery, brewery and textile factory.

Zn effluent concentrations were particularly high in the steel factory effluents, for both campaigns, with higher median concentrations in 2014, when the steel galvanizing processing was expanded (Table 2.1.). The Zn concentrations in the steel factory effluents far exceeded the USEPA and EMoI guidelines during the two campaigns (Table 2.1.). The loading of Zn

Table 2.1. Estimates of EC, pH, and of effluent concentrations and guidelines (µg/L), as well as standard errors (µg/L), effluent discharges (L/s) and daily loadings (g/day) of heavy metals in the five factories' effluents, during the first (C1) and second campaign (C2), from June–September 2013 and 2014, respectively. For the effluent loadings, the "direct median loading method" was used, n = 8

Factory Campaign (n = 8)		Steel C1	C2	Textile C1	C2	Tannery C1	C2	Meat processing C1	C2	Brewery C1	C2
EC (µS/cm)	Median	5730	3800	932	760	710	4470	1480	1590	920	1130
	Mean	14,400	4000	920	800	2200	5200	920	1200	2100	1800
	Maximum	78,000	7460	1190	1010	10,570	12,280	1170	1740	7100	3070
	Minimum	1430	620	730	480	450	800	560	740	720	1,070
	Standard error	920	790	54	63	1240	1500	77	116	731	247
pH	Median	6.1	5.5	10.3	8.2	7.8	7.4	8.2	7.2	11.1	11.2
	Maximum	6.1	10.9	10.2	8.8	7.8	8.1	8.2	8.2	11.8	11.4
	Minimum	0.4	2.2	7.5	7.7	7.4	5.6	6.7	7.1	5.2	6.9
	Standard error	0.7	1.1	0.4	0.1	0.0	0.4	0.4	0.1	0.7	1.1
Cr	Median (µg/L)	89	17	4.1	3.1	6.1	26,800	2.2	9	10	40
	Mean (µg/L)	150	32	4.1	45	22	33,270	2.1	60	8	36
	Maximum (µg/L)	485	85	4.9	297	131	64,600	2.1	215	16	77
	Minimum (µg/L)	2.1	1.1	2.2	2.1	2.3	813	2.3	1.1	2.1	2.9
	Standard error (µg/L)	60	11	0.7	36	17	7,850	0	34	2	8
	USEPA guideline[a] (µg/L)	1300	1300	N.A.[b]	N.A.	12,000	12,000	N.A.	N.A.	N.A.	N.A.
	EMoI guideline[c] (µg/L)	1000	1000	1000	1000	2000	2000	N.A.	N.A.	N.A.	N.A.
	Mean effluent (L/s)	1.7	2.2	15.4	16.5	6.8	8.4	11	8.8	8.2	21
	Loadings (g/day)	11	4	3	4	2.5	18,500	1.1	6	4	40
Cu	Median (µg/L)	65.2	99	14	6.9	11	15	9.1	3.1	25	26
	Mean (µg/L)	125	137	58	13	125	22	31	6.8	111	43
	Maximum (µg/L)	440	340	290	50	290	85	160	20	290	200
	Minimum (µg/L)	8.5	0.1	3.5	0.1	8.1	0.1	2.5	0.1	4.9	1.4
	Standard error (µg/L)	45	54	34	6	51	0	10	20	3	47
	USEPA guideline (µg/L)	1300	1300	N.A.	N.A.	N.A.	N.A.	N.A.	N.A.	N.A.	N.A.
	EMoI guideline (µg/L)	2000	2000	2000	2000	N.A.	N.A.	N.A.	N.A.	N.A.	N.A.
	Mean effluent (L/s)	1.7	2.2	15.4	16.5	6.8	8.4	11	8.8	8.2	21
	Loadings (g/day)	6	20	22	9	6.3	10	5	3	17	29
Zn	Median (µg/L)	60,040	155,750	120	110	90	280	110	140	150	210
	Mean (µg/L)	170,000	172,600	200	230	980	390	160	150	210	220
	Maximum (µg/L)	662,700	450,700	7190	640	7190	1250	180	330	720	440
	Minimum (µg/L)	14,100	14,150	26	29	26	130	25	44	20	68
	Standard error (µg/L)	87,800	50,110	76	85	887	0	125	43	33	76
	USEPA guideline (µg/L)	3500	3500	N.A.	N.A.	N.A.	N.A.	N.A.	N.A.	N.A.	N.A.
	EMoI guideline (µg/L)	5,000	5,000	5,000	5,000	N.A.	N.A.	N.A.	N.A.	N.A.	N.A.
	Mean effluent (L/s)	1.7	2.2	15.4	16.5	6.8	8.4	11	8.8	8.2	21
	Loadings (g/day)	4950	17,300	207	160	54	210	47	100	114	280
Pb	Median (µg/L)	5.1	8.2	2.9	1.1	2.1	2.1	2.9	1.1	5.9	1.1
	Mean (µg/L)	16	22	4.1	1.7	3.1	130	3.2	2.1	4.9	2.1
	Maximum (µg/L)	43	66	7.1	4.1	3.9	1670	4.1	2.9	8.1	2.9
	Minimum (µg/L)	2.1	0.6	2.1	0.6	2.1	0.6	2.1	0.6	2.1	0.6
	Standard error (µg/L)	5.7	9.5	0.7	0.7	0.3	0.0	233	0.2	0.2	0.7
	USEPA guideline (µg/L)	120	120	N.A.	N.A.	N.A.	N.A.	N.A.	N.A.	N.A.	N.A.
	EMoI guideline (µg/L)	500	500	500	500	N.A.	N.A.	N.A.	N.A.	N.A.	N.A.
	Mean effluent (L/s)	1.7	2.2	15.4	16.5	6.8	8.4	11	8.8	8.2	21
	Loadings (g/day)	1	1.3	3	1	1	4	1	0.6	3	2

[a] USEPA (2014)
[b] N.A. not available; no guideline concentration is given, [c] EMoI (2014)

from the textile factory was also relatively high, though less marked, during the C1 campaign (no guidelines are set for Zn in textile and tannery effluents; Table 2.1.).

The mean Pb concentrations and loadings increased, largely from tannery effluents, during the C2-campaign (Table 2.1.). Although, no guidelines are set for Pb in tannery effluents, maximum Pb values exceeded the guideline values set for Pb in the steel processing and textile industries effluents (Table 2.1.). The expansion of the steel processing factory in 2014 may have resulted in increased Pb concentrations in the effluents during the C2-campaign.

Metals in the effluent mixing zones of the Leyole and Worka rivers

The upstream catchments of the Leyole and Worka rivers are largely under agricultural use and in the upper parts, stations LD1 and WD1 were considered as "background" stations to compare with the concentrations of heavy metals downstream. However, at LD2, some increased Cr, Cu, and Zn concentrations were observed (Table 2.2.). In contrast, at WD1, the median concentrations of all heavy metals were lower than at WD2.

EC values were somewhat higher in the effluents mixing zones for the tannery, meat processing and brewery factories, similar to the earlier mentioned EC values in their effluents (Table 2.1.). In contrast, though EC was highest in the steel processing effluents, these effluents were largely diluted with river water and, therefore, no increased EC values in the steel factory's mixing zone were observed compared with the other mixing zones. Similarly, no marked pH effects were observed in the effluent mixing zones, except for high pH values down-stream of the brewery, in 2014 (Table 2.2.). The effect of the factories effluent discharges on the metal concentrations in the downstream river water was examined in the effluents mixing zones of the Leyole and Worka rivers (Figure 2.2.; Table 2.2.). The Cr concentrations were highest at LD4 for the C2 campaign (median: 2660 µg Cr/L), similar to the tannery factory effluent itself (factory not operational during 2013 Campaign; Table 2.1.). The median Cr concentration at the tannery effluent mixing zone was increased by a factor 52 (2660 vs. 51 µg Cr/L) compared with the nearest upstream station for the C2 campaign (Table 2.2.). Although Cr concentrations at LD5 were still relatively high, there was a marked decrease compared with LD4, probably due to increased dilution from numerous small streams flowing into the Leyole River between

Table 2.2. Estimates of EC, pH, and the metal concentrations (µg/L), flow rates (L/s) and loadings (g/day) for the industrial effluents mixing zones (M.z.) of the Leyole and Worka rivers. The flow rates (in italic) at LD 2–4 were estimated by interpolation, taking the average of flow rates at LD1 and LD5. The loadings were calculated as the product of median concentrations and flow rates of the rivers

Station		LD1		LD2		LD3		LD4		LD5		WD1		WD2	
Campaigns		C1	C2	C1	C2	C1	C2	C1	C2	C1	C2	C1	C2	C1	C2
EC (µS/cm)	Median	620	530	570	460	750	550	750	980	760	850	430	340	680	1240
	Mean	540	490	540	420	700	550	740	1050	770	850	400	350	700	1280
	Maximum	718	685	617	574	1080	650	1010	1480	1110	1260	480	470	990	2850
	Minimum	200	150	280	180	520	400	420	710	440	290	290	240	430	570
	Standard error	66	65	38	43	62	32	66	105	69	113	24	27	71	241
pH	Median	7.5	8.0	8.1	8.3	8.3	8.1	7.8	7.9	7.6	7.6	8.1	8.4	6.3	9.5
	Maximum	8.3	8.2	8.5	8.7	8.8	8.5	8.2	7.9	8.5	7.9	8.5	8.7	9.5	11.2
	Minimum	7.3	7.2	7.2	7.6	7.9	7.6	7.1	7.4	7.4	7.3	6.4	8.0	4.4	6.9
	Standard error	0.8	0.13	0.8	0.13	0.89	0.13	0.83	0.07	0.84	0.09	0.83	0.1	0.74	0.58
Cr	Median (µg/L)	3.9	2.1	12	6.1	7.9	51	9.1	2660	8.9	280	2.1	2.1	7.1	38
	Mean (µg/L)	3	440	11	380	6.9	230	9	6880	11	4280	3.1	37	7.9	30
	Maximum (µg/L)	21	2690	44	2160	25	1130	15	25,900	16	18,250	4.9	154	13	73
	Minimum (µg/L)	1.9	1.1	2.1	0.7	2.1	0.7	1.9	206	2.1	26	2.1	1.2	2.1	2.1
	Standard error (µg/L)	4.1	330	5.1	260	3.1	140	8.9	3360	6.1	2580	0.1	22	1.1	9.1
	Mean river flows (L/s)	98	184	120	240	135	277	138	287	142	296	360	1320	360	1,320
	Loadings (g/day)	34	32	124	124	93	1220	110	66,000	110	7260	62	228	218	4330
Cu	Median (µg/L)	23	0.4	17	14	63	41	10	21	14	27	8	0.2	13	33
	Mean (µg/L)	80	300	83	270	100	160	41	85	65	190	51	34	73	350
	Maximum (µg/L)	303	1900	248	1540	250	830	250	360	270	1180	270	150	270	2450
	Minimum (µg/L)	3.1	0.1	6.9	0.1	4.1	0.1	2.9	0.1	3.1	0.1	2.1	0.1	3.1	0.1
	Standard error (µg/L)	37	240	36	190	37	100	30	45	33	140	33	22	35	300
	Mean river flows (L/s)	98	180	120	240	130	280	140	290	140	300	360	1,320	360	1,320
	Loadings (g/day)	195	6	176	290	735	980	119	521	172	691	249	23	404	3,760
Zn	Median (µg/L)	72	110	95	520	71	187	30	205	81	214	41	137	106	194
	Mean (µg/L)	77	110	109	886	91	525	52	384	127	528	67	151	194	175
	Maximum (µg/L)	126	3310	367	2780	218	1600	131	1050	611	2120	143	338	855	278
	Minimum (µg/L)	26	16	29	9.1	54	34	15	67	15	25	8.9	12	14	46
	Standard error (µg/L)	15	402	37	365	21	209	17	127	65	250	19	45	92	29
	Mean river flows (L/s)	98	184	120	240	135	277	138	287	142	296	360	1320	360	1320
	Loadings (g/day)	610	1750	985	10,800	828	4480	358	5080	994	5470	1280	15,630	3300	22,130
Pb	Median (µg/L)	2.1	1.1	2.9	1.1	2.9	3.1	3.9	5.1	3.1	0.8	2.1	1.1	3.9	1.1
	Mean (µg/L)	1.1	11	1.1	9.9	1.1	8.1	0.4	128	1.1	7.9	3.1	2.1	2.1	1.1
	Maximum (µg/L)	4.9	70	6.1	60	4.9	34	4.1	980	4.1	44	3.9	7.1	4.9	5.1
	Minimum (µg/L)	2.1	1.1	2.1	1.1	1.9	1.1	2.1	0.6	2.1	0.6	2.1	1.1	2.1	1.1
	Standard error (µg/L)	0.4	8	0.7	7	0.4	3.9	4.1	121	0.4	5.1	0.3	0.7	0.2	0.4
	Mean river flows (L/s)	98	184	120	240	135	277	138	287	142	296	360	1320	360	1320
	Loadings (g/day)	17	16	31	21	35	72	48	124	37	20	62	114	124	114

LD4 and LD5. In line with the observed effluent Cu concentrations (Table 2.1.), no markedly increased Cu concentrations were observed in the mixing zones of the Leyole River except for relatively high median Cu concentrations during campaign C1 at LD3 in the textile effluent mixing zone (Table 2.2.).

In the Worka river, the median Cu concentration for both the C1- and C2-campaigns was higher at WD2, the mixing zone of the brewery effluent, than at WD1 (Table 2.2.). No comparable increases at WD2 were observed for the other heavy metals. Consistent with Zn in the steel processing factory effluent (Table 2.1.), highest Zn concentrations were found at the effluent mixing zone (LD2) with medians of 95µg Zn/L and 521µg Zn/L during C1 and C2, respectively. Just as for Cr, the Zn river concentration decreased again at LD3 (textile effluent mixing zone), reflecting a dilution effect from numerous water inflows into the river (Table 2. 2.).

The median and mean of Pb concentrations in the effluent mixing zones for both the Leyole and Worka rivers were both quite low, comparable with Pb values at LD1. Finally we tried to match, for both the Leyole and Work rivers, the metal loadings (g/day) as calculated from the factories' discharges (Table 2.1.), with those calculated at the effluent mixing zones, as products of median metal river concentrations with river discharges (Table 2.2.). Since the Leyole river discharges were not measured between LD1 and LD5, we assumed, by linear interpolation based on the distances between stations, that river discharges at LD2, 3 and 4 amounted to, respectively: 120, 135 and 138 L/s, for campaign C1, and 240, 277, and 287 L/s, for C2 (Table2.2.). Important results for these comparisons were, apart from extreme Zn and Cr loadings, rarely found (see later).

2.4 Discussion

2.4.1 Industrial Development and Pollution Management in the Kombolcha Industrial Zone

In 2010, the Ethiopian government implemented a 5 year Growth and Transformation Plan (GTP) through industrial growth and development. To realize industrial growth, the government identified five suitable sites (EMoI, 2014). Here collaboration takes place with the International Development Association of the World Bank to implement the Industrial Development Zones Projects (IDZPs). The GTP is currently in the second (GTP II) of three phases in the planned transition as national structural changes from an agriculturally to industrially-led economy. After the structural changes have been effected, the government envisages, in the third phase (GTP III), to attain a middle- income state (per-capita income of 1,200 USD per year) by the end of 2025.

Kombolcha, one of the five national IDZPs sites, is considered an ideal location because of its intermediate location for domestic markets exports via the Djibouti port (Figure 2.1.a). The city administration has allocated 1100 ha of land for industrialization (Mesfin, 2012). Labour-intensive manufacturing industries are a priority area for the industrialization process. Abundant cheap labour force and opportunity for duty-free exports to the USA has stimulated international investors to engage in medium to large-scale manufacturing industries. Existing factories are also expanding. The BGI-brewery, and the Kombolcha textile and steel processing factories have recently undertaken major expansions. The Ethiopian Industrial Development Zone Corporation (EIDZC) is responsible for planning, implementation and supervision of environmental issues for the industrial projects. The regional and city environmental institutions are charged to ensure good environmental management of the projects. For the Kombolcha IZDP, the Amhara Regional Environmental Authority is responsible for coordinating the industrial pollution regulations. At local level, the Kombolcha Bureau of Environmental Protection, Land Administration and Use (EPLAU) is responsible for monitoring industrial pollution and evaluating compliance with environmental requirements. The five factories examined in this study are located close to each other, with the new industries constructed in nearby areas. This will obviously increase pollution risks into receiving rivers. However, up until now we found no report dealing with environmental considerations for the Kombolcha IDZPs implementation, nor assessment studies on the carrying capacity of the surrounding environment with respect to expected industrial pollution.

2.4.2 Industrial Effluents and Metals Pollution in the Kombolcha Industrial Zone

In the Kombolcha industrial zone, effluents discharged by each factory are managed independently. In spite of the close proximity of the factories, we observed no joint efforts by the factories to manage waste disposal. Currently no treatment facilities are present for the brewery (Table 2.3.).For the other four industries, treatment takes place in lagoons or retaining ponds, but these facilities are quite old and designed to treat organic and sediment wastes only, rather than metal pollutants. According to the Environmental Pollution Control Proclamation of Ethiopia, all factories in the Kombolcha industrial zone are required to comply with national effluent emission standards, as each factory falls in the category for which emission standards are developed. Governmental environmental protection institutions both at the federal and regional levels coordinate the inspection of emission from the factories (for details, see next section) (Afework et al., 2010; EEPA, 2010; FDRE, 2002a).

Table 2.3. Expected effluent compositions for the five Kombolcha industries, type of treatment facility, and emission monitoring, as observed in 2015

Factory	Expected effluent composition	Treatment facility
Steel processing	toxics: As, CN, Cr, Cd, Cu, Fe, Hg, Pb, Zn; non-toxic: Fe^{3+}, Ca^{2+}, Mg^{2+}, Mn^{2+}.	Retaining ponds
Textile	Acid and alkaline, disinfectants: Cl_2, H_2O_2, formalin, phenol	Facultative lagoons
Tannery	Cr and organic wastes (i.e. Bio- oxidizables (BOD))	Anaerobic lagoons
Meat processing	Organic wastes, suspended solids, and BOD, nutrients (P, N)	Anaerobic lagoons
Brewery	organic wastes, suspended solids, BOD, nutrients (P, N)	No treatment facility

Our study in 2013 and 2014 could only take place in the rainy seasons, when the effluents encountered higher dilutions owing to increased flows of the rivers. In the dry seasons, reduced dilution will lead to more serious pollution. The chromium in the tannery effluents comes from the commonly used chromium salt $Cr_2 (SO_4)_3 12(H_2O)$, for tannery processes (Akan et al., 2007; Pawlikowski et al., 2006). The low Cr concentration in the tannery effluents during the 2013 campaign (Table 2.1.) can be attributed to the very low tanning production that year. According to the factory manager (Ali Mohammed, personal communication; 1 August, 2013), the factory process was then strictly limited to the preparatory steps before tanning, without the vegetal and chrome tanning processes involving Cr. In 2014, we found Cr concentrations as high as 64,600 µg/L in the tannery effluents exceeding both the USEPA and EMoI guidelines (Table 2.1.). Similar observations are reported from other developing, and Sub-Saharan countries (Table 2.4.).

Though the dilution factors of the Leyole river increased during C2 compared with C1 (see section *"Discharges of the Leyole and Worka rivers"*), Cr increased in the tannery effluent mixing zone, by a factor 51, compared with the nearest upstream station (Table 2.1.) during full tannery production. With a comparable factory production capacity, Gebrekidan et al.

(2009) and Katiyar (2011) reported enhanced river Cr concentrations downstream of the tannery effluent, at both high and low flows.

Table 2.4. Metals discharges from selected factories in Sub-Saharan and other developing countries

Factory effluent	Metals	Concentration (µg/L)	Country	Reference
Tannery	Cr	23,020	Kenya	Mwinyihija et al. (2006)
		10,820	Ethiopia	Gebrekidan et al. (2009)
		5790	Nigeria	Emmanuel and Adepeju (2015)
		3540	Ethiopia	Ayalew and Assefa (2014)
		264,000	Uganda	Oguttu et al. (2008)
		811,410	Morocco	Ilou et al. (2014)
		95,000	India	Ganesh et al. (2006)
		77,000	Albania	Floqi et al. (2007)
		5, 420,000	Bangladesh	Hashem et al. (2015)
	Pb	1060–1920	Nigeria	Akan et al. (2007)
		2870–3100	Nigeria	Emmanuel and Adepeju (2015)
		760	Morocco	Ilou et al. (2014)
		1970	Pakistan	Tariq et al. (2006)
	Zn	5520	Nigeria	Adakole and Abolude (2009)
Steel processing		2900	Bangladesh	Ahmed et al. (2012)
		168,150	Romania	Alexa (2013)
		498,500	India	Majumdar (2007)
Textile	Cu	5140	Nigeria	Yusuff and Sonibare (2004)
		2200–4500	Nigeria	Ohioma et al. (2009)
		1090	Pakistan	Sial et al. (2006)
		1700	Pakistan	(Manzoor et al., 2006)

In the effluents of the tannery factory, we also observed Pb peaks of up to 1670 µg/L when the factory was fully operational (Table 2.1.). This is probably connected to the use of Pb in the finished and unfinished trim process in post tanning operation (Akan et al., 2007). For comparable cases of 15 tanneries in Pakistan (Tariq et al., 2006) and two in Nigeria (Akan et al., 2007), high Pb contents were observed as well, often leading to violation of water quality guidelines. Aklilu (2013) showed that Pb consistently exceeded FAO irrigation quality guidelines downstream of the tannery effluent mixing point of a river in Ethiopia.

The major operation of the steel processing factory is to heat and galvanize the steel products with zinc coats, leading to high Zn concentrations in the factory's effluents (Rungnapa et al., 2010). Zn in the Kombolcha factory effluents often exceeded both the USEPA and EMoI guidelines (Table 2.1.). Similarly, very high Zn concentrations, up to 500 mg Zn/L have been recorded in effluents from steel processing factories elsewhere (Table 2.4.). Rungnapa et al.

(2010) found that the hot dip-galvanized process resulted in major eco-toxicity. In our study, though the steel effluent was rich in Zn (Table 2.1.), we found remarkably low Zn contents in the effluent mixing zone of the Leyole River (Table 2.2.). This is due to the large dilution effect of the steel processing effluents into the Leyole River for the C1 and C2 campaigns, by factors of 70 and 109 (i.e. based on the effluent and river flows data (Tables 2.1, 2.2), respectively.

Cu was higher in the steel factory effluents and we also observed similar variations of Cu concentrations in the textile and steel effluent mixing zones (Table 2. 2.). Many studies reported high Cu concentrations in the effluents of textile factories, largely related to the colouring of the fabrics (Ghaly et al., 2014; Dwina et al., 2010; Sial et al., 2006). We hardly found elevated Cu concentrations in the textile effluents (Table 2.1.), though the Cu concentrations were higher in the textile effluent mixing zone than for both the textile effluents and the other effluent mixing zones in the rivers (Table 2.2.). In July 2014, a visit to the textile factory, indicated that treatment of effluent waste comprised only a facultative lagoon constructed to treat organic wastes rather than dissolved heavy metals. Depending on the chemical products used for dyeing and the textile wet processing which is done at different times, pollutants in the effluent vary with time (Choudhury, 2006) and thus, the monitoring interval for this study (i.e. 2 weeks) may also not have been adequate to capture the variations of Cu concentrations in the effluent. In general, the quantity and quality of industrial effluents vary with discharges, operation start-ups and shutdowns, and working hours distributions (Henze and Comeau, 2008). While more frequent and preferably continuous sampling would have been desirable, this was not possible within the resource of the project.

Generally, we found similar trends for the metals concentrations in the factories' effluents compared with those in the effluent mixing zones of the rivers (Tables 2.1, 2.2). Cr and Zn concentrations in the effluents of, especially, the tannery and steel processing factories provide clear evidence of pollution. Cu and Pb showed similar trends, though less frequent compared with Cr and Zn. The effluents from the brewery and the meat processing factories showed relatively low metal concentrations (Table 2.1.). These effluents primarily comprise biodegradable/non-degradable organics and suspended solids, as well as nutrients such as ammonia, nitrate and phosphate (Inyang et al., 2012). More details are presented in Chapter 4.

For each metal, we found large differences in estimated loadings (g/day) from the effluents and the mixing zones (Tables 2.2, 2.3). Even though the frequency of monitoring the effluents and mixing zones were synchronized in this study, this was not always the case for the measurement of the effluent and river flows. For the Leyole river, the relative large influence of metal loadings from the upstream "background station" LD1 (Table 2.2.) distort comparisons, but not the overall conclusions concerting high industrial driven pollution. The impact of this station, e.g. as a source of diffuse metal loadings, as well as effects of the metal loadings on the river and sediment qualities in the region, will be discussed elsewhere (Zinabu et al., unpublished).

The large differences between the effluent and river water discharges (for Leyole river by a factor of 9–109; Worka river: 44–63) (Tables 2.1, 2.2), with, at the same time, relatively low metal river water concentrations, was an additional factor for the large loading differences. Even for the extremely high Cr discharges from the tannery during campaign C2, there was a factor 3.6 (66,000/18,500) difference between the estimated effluent and stream loadings (Tables 2.1, 2.2). For the Zn discharges from the steel factory during campaign C2, relatively less difference was found with a factor of 1.6 (17,300/10,800) between the two estimated loadings (Tables 2.1, 2.2). Effluent impacts in the effluent mixing zone of receiving rivers is generally affected by variations in river flows and geomorphology of the river flows receiving the effluents, as well as effluent density and temperature differences between effluents and receiving water (Alonso et al., 2016; Schnurbusch, 2000). These factors likely affected pollutant transport in the mixing zones in the Leyole and Worka rivers. Finally, since the rainfall distribution in the Kombolcha catchments is erratic and river bank erosion is evident over large parts of the river, mostly because of overgrazing and lack of erosion protection measures, both the earlier defined "Zone of initial dilution" (ZID) and chronic mixing zone (impact zone) likely vary over time and space. Thus, the 5 m long mixing zone selected for our study will not always have represented the actual mixing zones of the effluents.

In sub-Saharan countries, estimating pollutants loadings from factories using frequent monitoring over long duration may be difficult, both technically and cost-wise. Infrastructure for monitoring works are generally limited and water quality information is scant (Kamiya et al., 2008; Driscoll et al., 2003). Thus other, more economical methods giving comparable results must be chosen. Relating factories' specific heavy metals loadings (i.e. emission factors) to the associated activities resulting into the metal discharges, may be more appropriate than estimating loadings based on frequent monitoring of pollutant concentrations and measurement of flows in rivers (USEPA, 2014). This holds especially in case of easier and less-cost activities, and is more useful in areas where monitoring infrastructures are challenging and water sampling is problematic due to low hydrological flows, like the Leyole and Worka rivers (Figure 2.3.).

It is important to note that high metal concentrations in the upstream parts of both the Leyole and Worka rivers showed the presence of sources of heavy metals other than the factories listed in this study. The existing landfills and intensive agricultural activities in the area are likely sources of these heavy metals, and additional study is needed to assess their inputs. Comparing the metal loadings at LD1 and LD2 (Table 2.2.), we estimated that, on average, the former contributed 47% to the latter loadings, with minimum and maximum values of 2% (Cu; campaign C2) and >100% (Cu; campaign C1), respectively.

2.4.3 Industrial Pollution Control Policy and Implementation in Ethiopia

The Ethiopian Federal government has already formulated a series of environmental proclamations pertinent to sustainable development, including the proclamation of the Environmental protection organs (FDRE, 2002b), the Environmental pollution control proclamation (FDRE, 2002a), the Environmental Impact Assessment (EIA) proclamations policy (FDRE, 2002a) and the Water resources and management proclamation (EMoWR, 2004a). Empowered by the Environmental pollution control Proclamation No. 300/2002, the EEPA (Ethiopian Environmental Protection Authority) has formulated practicable emission standards that are generally required to be fulfilled by eight categories of factories liable to it (EEPA, 2010), but there are several weaknesses to the Ethiopian regulatory structure for pollution control (Table 2.5.). The factories are responsible not to exceed emission standards and to dispose effluents in an environmentally sound manner (Article 4 (1)). A factory that discharges a potentially dangerous pollutant is required to immediately notify the competent environmental authority (Article 4 (4)). Penalty for violating the regulations are referred as criminal code that is elaborated with Clauses (Article 14 and Part Five (Offences and Penalties, Articles 12 to 17)). EEPA is also in charge of supporting technical guidance for Environmental

Institutions at regional and sectorial levels; the regional states in turn transfer tasks to local levels. Subsequently, EEPA has now evolved into the *Ministry of Environment, Forest and Climate change*, but it is not clear yet whether the tasks will be changed or not. Here, we assume that EEPA will only be promoted administratively to ministerial level and that the tasks will remain unchanged. The emission standards set by EEPA are only focused on a limited number of pollutants. In principle, the EEPA guidelines could be technology-based (i.e. best available techniques (BAT)) or environment-based (i.e. environmental quality objectives or standards (EQOs)). Both methods generally demand detailed technological, economic and environmental considerations (OECD, 1999). Looking at the guidelines classification scheme based on eight industrial categories in Ethiopia (Table 2.5.), it is clear that EEPA uses BAT permits as precautionary measures.

Estimating combined loads of diffuse and point-source pollutants into the Borkena River, Ethiopia

Table 2.5. Description of Ethiopian pollution regulation and control components and, analysis of strengths, weakness and possible solutions

Issue	Industrial effluent pollution			
Pollution regulation and Control				
Regulatory structures	Federal level (EEPA), Regional level (REPA), Local level (Kombolcha Bureau of Environmental Protection, Land Administration and Use (EPLAU))			
Regulatory organs	Federal environmental institutions and the Council (Ethiopian Ministry of Environment, Forest and Climate change), Regional environmental institutions, Sectorial environmental institutions			
Control and command	Emission standards (limits of effluent quality discharge into water for eight categories of industries including (EEPA 2010)): Tanning and the production of leather goods; The manufacture of textiles; Extraction of mineral ores, the production of metals and metal products; The manufacture of cement and cement products; Preservation of woods and manufacture of wood products including furniture; The production of pulp, paper and paper products and; The manufacture and formulation of chemical products including pesticides.			
Strengths	Manifestation of Ethiopian Environmental Policy Formulation of laws and regulation to control industrial pollution (proclamations of the environmental protection organs; Environmental Pollution Control proclamation; the Environmental Impact Assessment (EIA) proclamations; and the Water Resources and Management proclamation)			
Weaknesses	Priority given to development over environmental protection	Lack of regulatory oversight relating to EIA	Reliance on use of effluent limits	Absence of any requirement to monitor or report for compliance of effluent limits
Source of weaknesses	• Lack of awareness and political commitments to environmental protection • Absence of clear links between development objectives and environmental protection • Foreign investor indifference to environmental protection	• Lack of effective rules and legal enforcement for EIA • Lack of environmental protection awareness by EIA licensing bodies • Absence of political commitment • Lack of communication among EIA regulatory institutions	• Lack of financial and technical resources by concerned institutions • Lack of economic incentive • Limited monitoring infrastructure for effluent receiving environments such as rivers • Lack of clear protection guidelines to effluent mixing environments	• Absence of rules for clear monitoring schemes for industrial pollutants • Limited professional, technical/finance capacity • Absence of technology standards to control pollution by industries • Lack of enforcement to compliance emission guidelines • Lack of transparency (for public use) in monitoring records
Possible solutions	• Awareness raising of decision makers in environmental protection • Prioritizing sustainable development in policy formulation and guidance	• Reformulating clear rules and strict implementation of EIA legal enforcement • Systematic use of EIA and coordinating the tasks of EIA regulatory institutions (e.g licensing organization and EEPA)	• Introducing economic incentives schemes i.e. collecting revenue from emission fees, taxes and subsidies • Expanding monitoring infrastructures • Developing effect based water quality guidelines after mixing of effluents in receiving water bodies	• Formulating clear rules for emission monitoring in industries • Developing technology based emission guidelines • Capacity building of emission controlling institutions • Strict follow up of legal enforcements • Public disclosure of available monitoring records • Development of environmental management systems linked with monitoring and reporting

As there are no guidelines after effluent mixing, it is impossible to clearly understand impacts of effluent emissions into receiving waters (Table 2.5.). To evaluate the Kombolcha industrial effluents, we used the more frequently updated USEPA guidelines. In 2014, the Ethiopian Ministry of Industry used these to prepare a draft Environmental guidelines framework financed by the World Bank (EMoI, 2014). At the moment, these EMoI guidelines are only intended to be used for the specific conditions of two industrial zones in the capital city, Addis Ababa. The guidelines include emission limits for pollutants from the same eight industrial categories, but are more recent than the more general EEPA guidelines (Table 2.5.).

According to the Environmental pollution control proclamation, both the federal and regional environmental protection authorities coordinate inspection of pollution sources to control violation (FDRE, 2002b). The regional state is also authorized to adopt emission permits and to control the more stringent industrial pollution areas. However, many studies indicate that downstream rivers are heavily polluted because of industrial wastes (Beyene et al., 2009a; Prabu, 2009). While industrialization has been growing fast for the past two decades, capacity within the regions and local environmental institutions have not kept pace with effective implementation of policy measures (World Bank, 2015).

Since 2008, EEPA has issued Directives to prevent environmental pollution. For licensing investment, EIA has been mandatory since 2003, but has been poorly implemented (CEPG, 2012; Demeke and Aklilu, 2008). Manufacturing industries are required to implement an environmental management plan and undertake environmental audit. However, new factories are often approved by licensing institutions (such as Ministry of Trade and Industry and Ministry of Mines and Energy) that regularly lack expertise, without the consent of EEPA and Regional Environmental Protection Authorities (REPA). Rather than carrying out EIA before the start of a project, in close communication with EEPA, licensing institutions often seem to rely on probable project outcomes with respect to monitoring and enforcement (Table 2.5.). An example is outlined by Getu (2009) and CEPG (2012) reported on several licensed floriculture industries severely polluting downstream aquatic resources by fertilizers and pesticides. In a similar case, Demeke and Aklilu (2008) pointed out how a foreign company was licensed, with no prior EIA, to work on biofuel projects on land located inside a wildlife sanctuary. In all cases, communication between the licensing institutions and EEPA/REPA was poor with failure to carry out EIA in a coordinated manner. EEPA has already formulated guidelines for environmental impact study reports in the eight industrial categories (Table 2.5.), but the EIA proclamation lacks clear understanding on the legal liability for improper implementation among the licensing institutions, environmental councils and sector bodies. EIA is often seen as a hindrance to development (CEPG, 2012; Demeke and Aklilu, 2008). Similarly, Ruffeis et al. (2010) indicated that in Ethiopia the investment proclamation tends to prevail over the EIA proclamation; allowing licenses without any obligations for an EIA.

In Ethiopia, given the limited financial capacity of, especially domestic, investors, financing Cleaner Production and waste treatment facilities are a high burden (CEPG, 2012; EEPA, 2010; Getu, 2009) and financial initiatives from government to support such investments are limited (Assefa, 2008). In developing countries, where factories are often traditional and small-scale (Jining and Yi, 2009), the rate of changing old technologies and adoption of environmentally sound ones is slow (Bertinelli et al., 2012; Rudi et al., 2012).

The implementation of the Ethiopian government industrialization plan, which stimulates growth of industries in specific zones throughout the country, would benefit from strong regulatory structures and pollutant monitoring. On the other hand, a frequent lack of respect by foreign investors towards multilateral environmental agreements (MEAs) and national environmental laws is a problem in many sub-Saharan countries (OECD, 2007). In our study area, the French Castel Group Company in Ethiopia, having a high awareness of the need for environmental protection as evinced from the Castel website (http://www.groupe-castel.com/en/environment/), has been operating for a number of years without effluent treatment facilities. During the study time the brewery indicated that a treatment system was pending, though it appears not yet to have been installed, nor was access to the factory allowed up the submission of this article. Finally, the public awareness on environmental protection is increasing in the Kombolcha, as evinced elsewhere in the sub-Saharan countries (OECD, 2013; Getu, 2009; Prabu, 2009). In our study, at the downstream of Worka river, farmers claimed that there has been declining crop production over the years because of the use of the brewery effluent mixed water for irrigation. Though the public can prosecute polluting industries violating environmental emission limits, it will be difficult to prove since pollution records, if kept at all, are stored centrally at the Federal and Regional environmental Institutions and hardly available for examination (Table 2.5). As indicated elsewhere in Ethiopia, public participation in EIA remains very limited in Kombolcha (Damtie and Bayou, 2008).

2.5 Conclusion

Over the years, the Kombolcha industrial zone has become attractive for domestic and foreign investors in, especially, manufacturing industries. The expansions of the existing and building of new industries has led to gradual pollutants increments and exemplify the challenges of industrial cities in sub-Saharan country's cities. Metals, especially Cr in tannery and Zn in steel processing factory effluents, were exceeding effluent emission quality guidelines. For Kombolcha, we suggest studies on the carrying capacity of the rivers that receive these industrial effluents. Further, a single centralized waste treatment facility used by multiple industries could be an efficient and cost-effective initiative. Though legislation on industrial emission permits, control and fines do exist, the capacity of the local and regional environmental protection institutions for industrial pollution strategies is very limited. The discrepancies on the institutional levels and disagreements between the environmental and investment policies and proclamations hampers successful enforcement of environmental pollution control. The non-adequate respect for (inter)national environmental agreements by (foreign) investors and the absence of governmental initiatives to support adoption of cleaner production techniques are other factors of importance. This study generally shows that the industrial investment path followed in the Kombolcha industrial zone is unsustainable with respect to environmental concerns for the rivers that receive the effluents.

To ensure effective implementation of environmental pollution control policies, the Ethiopian Federal and Regional governments could better facilitate local environmental controlling institutions with the required instrumentations and mechanisms for law enforcement. Investment and capacity building within local government's agencies can then provide long-term development of procedures and environmental protection. Ultimately, environmental protection is a social choice, and mechanisms that better involve all stakeholders, from local public to international investors, provide for the necessary dialogue and support of environmental regulations for both the region's and country's long-term sustainable development.

Chapter 3

Preventing sustainable development: policy and capacity gaps for monitoring heavy metals in riverine water and sediments within an industrialising catchment in Ethiopia

Publication to be based on this chapter:

Zinabu E, Kelderman P, van der Kwast J, Irvine K. 2018. Preventing sustainable development: policy and capacity gaps for monitoring heavy metals in riverine water and sediments within an industrialising catchment in Ethiopia. *In submission.*

Abstract

Managing water quality needs knowledge of pollutants, agreed standards of quality and a relevant policy framework that supports monitoring and regulation. In many settings, however, an effective policy framework or its application is absent. This paper reports on the assessment of heavy metals in the rivers within an industrialising catchment in Ethiopia, and the importance of improving policy and capacity for monitoring and management. For two sampling periods in 2013 and 2014, chromium (Cr), copper (Cu), zinc (Zn) and, lead (Pb) were monitored in water and sediments of the Leyole and Worka rivers in Kombolcha city, Ethiopia, and evaluated against international guidelines, and Ethiopian water protection policies. Chromium was high in the Leyole river water (median: 2660 µg/L) and sediments (maximum: 740 mg/kg), Cu concentrations in the river water was highest at the midstream part of the Leyole river (median: 63 µg/L), but maximum sediment content of 417 mg/kg was found upstream. Zn was highest in the upstream part of the Leyole river water (median 521 µg/L) and sediments (maximum: 36,600 mg/kg). Pb concentration was low in both rivers, but, relatively higher content (maximum: 3,640 mg/kg) found in the sediments in the upstream of the Leyole river. Cr showed similar pattern of enhanced concentration in the downstream part of the Leyole River, with Cu and Zn having significantly different concentrations between the monitoring periods. Except for Pb, the concentrations of all heavy metals surpassed the guidelines for aquatic life, human water supply, and irrigation and livestock water supply. All heavy metals exceeded guidelines for sediment quality for aquatic organisms. In Ethiopia, further development of water quality standards and an effective and locally relevant monitoring framework are needed. Current WHO guidelines used for drinking water quality are not designed for monitoring ecological health or account for local ambient water hardness. Poor technical and financial capabilities further hamper monitoring of rivers and sediments. Development of monitoring protocols and institutional capacities are both necessary and possible to support Ethiopia in its ambitions for increased industrialisation and agricultural intensification. Failure to do so presents high risks for both public and ecosystem health.

3.1 Introduction

Pollution of surface waters is often a particular problem in developing countries, especially where expansion of industry and intensification of agriculture are not matched with suitable water quality policies or, more commonly, their enforcement. Monitoring infrastructures can be limited and institutional set up lacking management and scientific capacity. Where monitoring has been done, access to information may not be forthcoming (Hove et al., 2013; Commission, 2011). Economic and financial pressures frequently dominate other concerns, with impact of pollutants on water bodies frequently neglected (Abbaspour, 2011).

Owing to their toxicity and persistence in aquatic systems, heavy metal pollution is a concern for public and ecosystem health (Yuan et al., 2011; Armitage et al., 2007). Heavy metals from industry and agricultural sources often end up in sediments (Islam et al., 2014; Su et al., 2013), where they may then be subject to remobilization in response to changes in geochemical ambient conditions such as pH and redox potential, themselves affected by hydrology, organic matter and sediment grain size (Kelderman, 2012; Yi et al., 2011). In sub-Saharan Africa, while many countries have adopted environmental quality standards, monitoring of heavy metals in water and sediment for legal compliance, or to guide industrial or land policies is often minimal (Chikanda, 2009). Ethiopia provides a striking example of how high pollution risk from multiple point sources and farming remains unmeasured. Despite Government awareness of potential impacts from pollution, there is limited action to protect human or ecosystem health (Akele et al., 2016; Aschale et al., 2016; Beyene et al., 2009b).

Ethiopia has endorsed several international conventions and agreements for water protection (EEPA 2010), is a signatory to the Sustainable Development Goals (SDGs) and has aligned the second Growth and Transformation Plan (GTP-II) to the sustainable development of the country (FDRE 2016). Nevertheless, rapid urbanization and industrialization continue to degrade surface waters, even though the Ethiopian environmental protection authority (EEPA), recently renamed "Ethiopian Ministry of Environment, Forest and Climate change (EMEFCC)", is legally responsible to formulate water quality policies that meet international standards, and establishing institutions to support that (EEPA, 2002; FDRE, 2002b). Policies regarding water protection are, however, limited to regulation of pollutants emission into waters through a "polluter pays" principle to be implemented by regional and local water bureaus, rather than a country-wide approach for preventive management. The city of Kombolcha in north-central Ethiopia is a typical example of a conurbation with extensive pollution from industrial and municipal discharges, as well as agricultural intensification (Zinabu et al., 2017a). While it is visually apparent that some stretches of the rivers running through the city are polluted, with recognition by city authorities of likely impact on, especially, human health, information on river water and sediment quality is scant. The rivers receive discharges from multiple industrial sources using heavy metals in their processing (Zinabu et al., 2017a), and surface water pollution is likely exacerbated by recurrent regional drought. While some monitoring of discharges occur by some factories, this is limited to variables such as BOD, COD, total suspended solids, and pH. No monitoring of heavy metals in discharges is done. While this would itself negate any assessment of compliance, in Ethiopia only WHO drinking-water quality guidelines are considered and there is an absence of national or local environmental standards for ecosystem health (EMoWIE, 2016). In this study, concentrations of the heavy metals: chromium (Cr), zinc (Zn), copper (Cu) and lead (Pb) were monitored over a two year period in the rivers running through Kombolcha, and heavy metal content compared with "a compendium of environmental quality benchmarks" (Macdonald et al., 2000a) and sediment quality standards as set out in the numerical Sediment Quality Guidelines (SQGs) (MacDonald et al., 2000b; USEPA, 1997a). Information from local government and review of policies in Ethiopia and the regions was used to make a more general overview of i) application of standards, ii) compliance, and iii) recommendation for improvement.

3.2 Materials and Methods

3.2.1 Study area description

Kombolcha, located in the North central part of Ethiopia (Figure 3.1.a, b), covers 125 km^2 comprising agricultural and forest land in the rural uplands and peri-urban area, and urbanized and industrial areas in lowland plains. Vertisol is the predominant soil type, with *Fluvisols* and *Cambisol* soil types common along the river bank and in the foothills, respectively (Zinabu, 2011). Our study was conducted in the Leyole and Worka rivers that receive point source discharges from four manufacturing industries and one beverage producer (Figure 3.1.c). These generally small rivers drain into the Borkena River (Figure 3.1.b), used for irrigation, rural water supply and wetland recharge; which in turn is a tributary of the larger Awash River.

3.2.2 Sample collection, preservation and analysis

River water monitoring and analysis

Water samples for the measurement of total dissolved Cr, Cu, Zn and Pb were collected by grab samples, using 100 mL polyethylene (PE) containers, that were filled just beneath the water surface at two confluence points of upstream sub-catchments main streams (LD1; WD1),

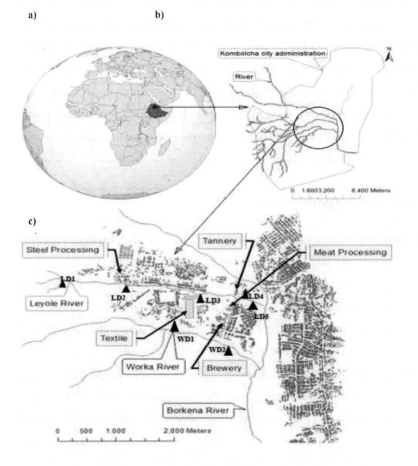

Figure 3.1. The location of the study area on the horn of Africa (a), in the Kombolcha city administration (b), and within the industrial zone areas and codes (c): (LD1 (confluence of three upstream tributaries and start of upstream Leyole river); LD2 (downstream of effluent discharge of steel processing factory in the Leyole river); LD3 (downstream of effluent discharge of textile in the Leyole river); LD4 (downstream of tannery effluent discharge in the Leyole river); LD5 (downstream of meat processing effluent discharge in the Leyole river); WD1(upstream Worka river); and WD2 (downstream of brewery effluent discharge in the Worka river))

and downstream of the factories effluent discharges, in the Leyole (LD2-5) and Worka river (WD2) (Figure 3.1.c). The locations were identical to the "effluent mixing zones" outlined by Zinabu et al. (2017a). Sampling was conducted during two periods in the rainy season from June to September 2013 and 2014 with, in total, 16 monitoring dates and 112 water samples for seven stations, following ISO 2003 recommendation. On each sampling occasion, pH and electrical conductivity (EC) were measured *in-situ* using a portable pH (WTW, pH340i) and EC (WTW, cond330i) meter, respectively (Zinabu et al., 2017a). Water samples, were acidified to a pH < 3 by adding 1 mL concentrated H_2SO_4, preventing heavy metal adsorption to the container wall. In keeping with ISO 5667-3, the samples were processed within 6 months (ISO, 2003), after transport to the IHE-Delft laboratory, The Netherlands. For processing, a 10 mL sample was filtered through Whatman GF-C glass microfiber filters (pore size 1.3 μm), and a final volume was made up to 100 mL using Milli-Q water. Heavy metal concentrations were measured using ICP-MS, XSERIES 2 IUS-MS. All analyses were done in accordance with "Standard Methods" (Rice et al., 2012).

Sediment monitoring and processing

River bed sediment samples were taken at six stations (LD2-5; WD1-2) on three occasions (Figure 3.1.): 15 June (M1), 15 July 2013 (M2) and on 15 July 2014 (M3) (in total 18 samples). Unlike the sampling in water, sediment samples were not taken from LD1. We assumed that measurements of heavy metals from this station represent background concentrations from geology and upstream agriculture lands. At each location samples were collected with a PE spoon from the upper 3 cm of sediment at four evenly spaced points across the width of the river. Individual samples were mixed, pooled, and stored in 250 mL PE beakers. The resulting approximate 50 grams of sediment was stored in the dark, transported to the Delft laboratory within four months, air-dried in a dark room within six weeks and, subsequently, analysed within 60 days.

To determine the sediment particle size distribution, samples were mixed and homogenised and Milli-Q water was then added (about 1:1 by volume) and the slurries mixed by stirring overnight at 150 rpm (IKA RW20) to dissipate clay aggregates. Slurries were then wet-sieved and shaken with warm tap water (Tritsch, mesh size 230 μm) to yield particle size fractions of: 63-125μm; 125-500μm, 500μm -1mm and 1-2 mm. The fraction > 2 mm was not taken as part of "sediment" (Håkanson and Jansson, 1983). The grain size fraction <63μm size was estimated by collecting all sediment particles passing through the 63 μm sieve after pre-settling overnight in a 25 L PE bucket, The particle fractions were then oven-dried at 70^0 C to constant weight. We determined the percentages of the five grain size fraction from dried fractions and calculated: a) median grain size (50% larger; 50% smaller) and b) sorting coefficient (S.C.)[1] (Håkanson and Jansson, 1983) The proportion of grain sizes at each station were computed using probability paper (Boggs, 2009). Both median grain size and sorting coefficient are expressed in dimensionless phi-units[2]. To determine organic matter (OM) content, approximately 1 g of each sediment sample was first heated in an oven at 70^0C till constant weight. After this, weight loss was determined by ignition at 520^0C for three hours (Rice et al., 2012). Triplicates of 0.5 g of the sediment fractions were taken from each grain size fraction and transferred into a Teflon tube; these were acid-digested with 10 mL 65% concentrated HNO_3. The fractions were dried and digested in a microwave oven (MARS 5) to determine heavy metal contents (mg/kg) using an Inductively Coupled Plasma-Mass Spectrometer (ICP-MS), Thermo Scientific X Series 2®, followed by dilution with Milli-Q water.

Statistical techniques

Since the relative standard deviations of the dataset of each heavy metal were high and the mean values affected by a few outliers (Table 3.1.), median values of the dataset were used to better represent the heavy metal concentrations at each station (Bartley et al., 2012; Pagano and Gauvreau, 2000). Two-way repeated measurements ANOVA (Analysis of Variance) was used to test temporal differences of heavy metal concentrations within and between the two season samplings. Analysis was done in R using "ANOVA" in a "CAR" (companion to applied regression) statistical package (R Core Team, 2015; Fox and Weisberg, 2011). Statistically significant difference between tested times frames was estimated by Kruskal Wallis one-way analysis of variance at $p \leq 0.05$, and Tukey HSD multiple comparisons.

---

[1] Equivalent with "standard deviation", $S.C = \left(\frac{1}{2}(grain\ size\ < 84\%_{of\ total}) - (grain\ size\ > 16\%_{of\ total}\right)$
Sediments with S.C. < 1.0 phi units are considered as "well sorted"; > 1.0 phi units as "poorly sorted" (Håkanson and Jansson 1983)
[2] $phi\ unit = -\log_2 grain\ size$; *e.g.* for grain size $= 0.63 = 2^{-4}$ mm; $phi\ size = 4.0\ phi\ units$

Quality assurance

All the reagents used were of analytical grade, and sample dilutions made with Milli-Q water. The precision of the analytical procedures, expressed as the relative standard deviation (RSD), was 5% – 10%. The ICP-MS measurements always had an RSD < 5% in the laboratory analyses. In all experiments, blanks were run; and during digestion of the sediment samples, two blanks of 65% HNO_3 and two standards of certified reference materials (sewage sludge amended soil-No 143 R ID 0827) were used. Calculated recovery percentages were: 100%, 103%, 104%, 105 %, for Cr, Cu, Zn and Pb, respectively.

Comparing water and sediment quality with environmental guidelines

Water hardness (measured as the concentration of $CaCO_3$ (mg L^{-1}) in the water) affects bioavailability of heavy metals (Pourkhabbaz et al., 2011; Besser et al., 2001). A previous study showed that the hardness of the Leyole and Worka rivers water is < 60 mg L^{-1}, and can be classified as "soft" waters (Zinabu 2011). We evaluated the compliance to environmental quality guidance for heavy metal concentration in the river water using under conditions of "soft" water Macdonald et al. (2000a). To examine the environmental quality of sediments in the rivers, we used the numerical Sediment Quality Guidelines (SQGs) (MacDonald et al., 2000b; USEPA, 1997a) for each of the three monitoring occasions. The sediment quality of each station was assessed using a *Threshold Effect Concentration* (*TEC*) and a *Probable Effect Concentration (PEC)*. TECs are the contents below which adverse effects on sediment-dwelling organisms are not expected, while PECs are the contents above which adverse effects are expected to occur frequently, and which may call for urgent remedial actions (MacDonald et al., 2000b; Swartz, 1999).

3.3 Results

3.3.1 Heavy metals in the Leyole and Worka rivers

Comparisons with water quality guidelines

Exceedance of one or more water quality guidelines were found for all of the four heavy metals in both rivers and years (Figure 3.2.). As expected, highest median concentrations of total dissolved Cr were found downstream of the tannery outflow (Table 3.1.), with dissolved Cr exceeding guideline limits in 2014, with a maximum of 25.9 mg/L recorded at LD4 in 2014 (Figure 3.2.a). Although median Cr concentrations decreased at LD5, the guideline limit for protection of livestock was still exceeded in 2014. Median Cr concentrations did not exceed guidelines at the other Leyole stations, though maximum Cr concentrations, especially for monitoring period C2, did. For WD 1 and 2, at the two stations in the Worka River, Cr concentrations were relatively low compared with the Leyole River. However, the Cr guideline limit for protection of aquatic life was exceeded here (Figure 3.2. a).

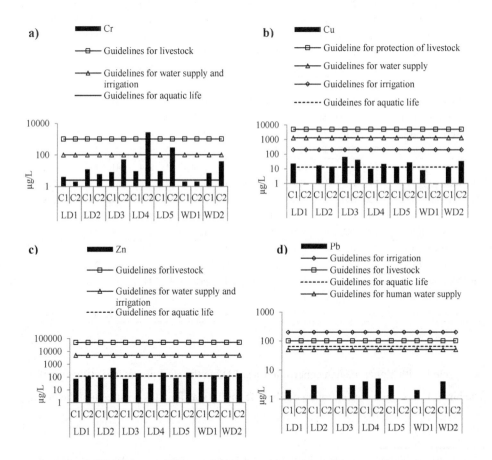

Figure 3. 2. Median heavy metal concentrations (µg/L) at the monitoring stations (see Figure 4.1.c) of the Leyole river and Worka river for the 2013 (C1) and 2014 (C2) monitoring periods; also the different water quality guidelines are presented[2]

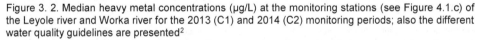

[2] Guidelines for protection of aquatic life in µg/L (for hardness ≤ 100 mg/L): Cr (2.5), Cu (13), Zn(120), Pb (65) USEPA. 1998. National recommended water quality criteria: Republication. Office of Water, United States Environmental Protection Agency Washington D.C.

For both monitoring periods, median concentrations of dissolved Cu were close to or exceeded the guideline limits for protection of aquatic life at all stations of the Leyole and Worka rivers (Figure 3. 2.b), with a highest median record of 0.06 mg/L at LD3 (in the textile effluent mix in the Leyole river) in 2013 (Table 3.1.). A similar pattern was observed for Zn (Figure 3.4.c),

Table 3.1. Estimates of heavy metal concentrations (µg/L) at stations LD1-5 in the Leyole River and WD1-2 in the Worka River

Station		LD1		LD2		LD3		LD4		LD5		WD1		WD2	
Monitoring periods (n=8)		C1	C2	C1	C2	C1	C2	C1	C2	C1	C2	C1	C2	C1	C2
	Median (µg/L)	4	2	12	6	8	51	9	2,660	9	284	2	2	7	38
	Mean (µg/L)	3	437	11	380	7	230	9	6,880	11	4,280	3	37	8	30
Cr	Maximum (µg/L)	21	2,690	44	2,160	25	1,130	15	25,900	16	18,250	5	154	13	73
	Minimum (µg/L)	2	1	2	0.7	2	0.7	2	206	2	26	2	1	2	2
	Standard error	4	330	5	260	3	140	9	3,360	6	2580	0	22	1	9
	Median (µg/L)	23	0.4	17	14	63	41	10	21	14	27	8	0.2	13	33
	Mean (µg/L)	80	305	83	268	101	155	41	85	65	188	51	34	73	354
Cu	Maximum (µg/L)	303	1,900	248	1,540	254	827	254	358	268	1,180	270	154	274	2,450
	Minimum (µg/L)	3	0.1	7	0.1	4	0.1	3	0.1	3	0.1	2	0.1	3	0.1
	Standard error	37	237	36	191	37	101	30	45	33	144	33	22	35	301
	Median (µg/L)	72	110	95	521	71	187	30	205	81	214	41	137	106	194
	Mean (µg/L)	77	110	109	886	91	525	52	384	127	528	67	151	194	175
Zn	Maximum (µg/L)	126	3,310	367	2,780	218	1,600	131	1,050	611	2,120	143	338	855	278
	Minimum (µg/L)	26	16	29	9	54	34	15	67	15	25	9	12	14	46
	Standard error	15	402	37	365	21	209	17	127	65	250	19	45	92	29
	Median (µg/L)	2	1	3	1	3	3	4	5	3	0.8	2	1	4	1
	Mean (µg/L)	1	11	1	10	1	8	0.4	128	1	8	3	2	2	1
Pb	Maximum (µg/L)	5	70	6	60	5	34	4	980	4	44	4	7	5	5
	Minimum (µg/L)	2	1	2	1	2	1	2	0.6	2	0.6	2	1	2	1
	Standard error	0.4	8	0.7	7	0.4	4	4	121	0.4	5	0.3	0.7	0.2	0.4

as expected, with higher median concentration at LD2 in the steel processing effluent outflow. The highest maximum concentration of Pb (2.45 mg/L) was found downstream of the brewery

Guidelines for protection of human health WHO/UNICEF. 2014. Progress on drinking water and sanitation: 2014 update. WHO, Geneva.in µg/L, for Cr(100), Cu (1300) and Pb(50), for Zn (5000) USEPA. 1986. Quality criteria for water. EPA, Washington D.C.
Guidelines for protection of irrigation in µg/L, for Cr (100), Nagpal, N., Pommen, L. & Swain, L. 1995. Approved and working criteria for water quality. Ministry of Environment, Victoria.; for Cu(200), Zn(5000); (soil pH > 6.5) and Pb (200) CCREM. 2001. Canadian water quality guidelines. Environmental Quality Guidelines Division, Ottawa.
Guideline for protection of livestock in µg/L, for Cr (1000) Nagpal, N., Pommen, L. & Swain, L. 1995. Approved and working criteria for water quality. Ministry of Environment, Victoria. for Cu(5000), Zn(50000) and Pb(100) CCREM. 2001. Canadian water quality guidelines. Environmental Quality Guidelines Division, Ottawa.

outflow in the Worka river (Table 3.1.), but Pb concentrations were generally well below all water quality guideline limits (Figure 3.2.d). Finally, contrary to expectation, high median Cu and Zn concentrations were recorded at station LD1, upstream of the industrial discharges in the Leyole River (Figure 3.2 a, b), possibly related to solid wastes of the factories in the vicinity (see later discussion for Zn).

Spatial and temporal variation of the heavy metals concentrations

Significantly different ($p \leq 0.05$, Kruskal Wallis test) concentration of Cr were found between stations LD1 and LD4 (Tukey HSD test, $p = 0.03$) (Table 3.2.). No significant difference in concentration between stations were found for the other three heavy metals concentrations.

Differences in temporal patterns were significant for Cu and Zn in the Leyole river (respectively, $p \leq 0.05$, 0.01) within and between years (Repeated-measures ANOVA) (Table 3.3.). In the Worka River, statistically significant differences were found only for Cr between years, and for Pb within years.

Table 3.2. Kruskal Wallis rank sum test of heavy metals among sampling stations along the Leyole river; d.f. = degrees of freedom; and Tukey HSD test for significantly varied heavy metal (p adj. = 0.05)

Test Group (LD1, LD2, LD3, LD4, LD5)	d.f.[a]	Kruskal-Walish Chi-squared	p-values
Cr	4	13	0.01*
Cu	4	1.2	0.88
Zn	4	2	0.82
Pb	4	1	0.89
Test heavy metal for HSD	Stations compared	p adj.	
	LD1 vs. LD2	0.9	
	LD1 vs. LD3	0.9	
	LD1 vs. LD4	0.03*	
	LD1 vs. LD5	0.07	
Cr	LD2 vs. LD3	1	
	LD2 vs. LD4	0.21	
	LD2 vs. LD5	0.38	
	LD3 vs. LD4	0.24	
	LD3 vs. LD5	0.42	
	LD4 vs. LD5	1	

d.f. = degree of freedom
*significant at $p \leq 0.05$

Table 3.3. Univariate Type III Repeated-Measures ANOVA test result for biweekly (BW) and monitoring period (MP) levels of mean heavy metals concentrations monitored at stations LD1-5 in the Leyole River and WD1-2 in the Worka River

River	Heavy Metals	Level of test	SS num[a]	d.f.[b]	error SS[c]	den d.f.[d]	F[e]	Pr (>F)[f]
Leyole river	Cr	MP	1.18E+08	1	1.45E+08	4	3.2	0.145
		BW	1.24E+08	7	3.84E+08	28	1.2	0.292
	Cu	MP	3.12E+05	1	1.01E+05	4	12	0.024 *
		BW	3.19E+06	7	1.42E+06	28	9.0	8.532e-06 ***
	Zn	MP	5.11E+06	1	5.31E+05	4	38	0.003 **
		BW	7.20E+06	7	4.97E+06	28	5.7	0.000 ***
	Pb	MP	7.96E+04	7	3.39E+05	28	0.9	0.492
		BW	1.87E+04	1	4.56E+04	4	1.6	0.270
Worka river	Cr	MP	5.17E+03	1	3.00E+00	1	1721	0.015 *
		BW	7.51E+03	7	4.42E+03	7	1.7	0.250
	Cu	MP	8.95E+04	1	2.24E+05	1	0.4	0.641
		BW	1.13E+06	7	1.32E+06	7	0.8	0.577
	Zn	MP	1.86E+02	1	1.27E+04	1	0.01	0.923
		BW	1.97E+05	7	2.05E+05	7	0.9	0.518
	Pb	MP	6.28E+00	1	2.33E+00	1	2.7	0.348
		BW	3.80E+01	7	6.70E+00	7	5.7	0.017 *

[a]sum of squares for numerator; [b]degree of freedom, [c]error sum of square, [d]denumerator degree of freedom, [e]F values, [f]p values: *** 0.001; ** 0.01; * 0.05

3.3.2 Heavy metals contents on the Leyole and Worka river sediments

Grain size distributions and organic matter contents

Grain size distributions were highly variable at the stations, both for the Leyole and Worka river stations with coarse sand (< 1.0 phi units) at LD3, LD5 and WD2, and at the other stations dominated by fine to very fine sand between 2.0 and 4.0 phi units) (Table 3.4.). The sediments at all stations were poorly sorted (sorting coefficient > 1.0 phi units). The OM contents of the sediments from the stations were rather uniform (Table 3.4.), but with a distinctively higher percentage (25%) at LD1 and LD2. Relatively, low OM content was observed in the Worka River compared with the Leyole River, with lowest content at downstream station (WD2).

Table 3.4. Overview of sediment characteristics in the dark shaded columns in the left (for median grain size (phi units), sorting coefficients (phi units), fine grain size distribution (<63μm %), and Organic Matter, OM (%)), and measured and corrected concentrations of Cr, Cu, Zn and Pb in the right part columns. Normalization is done with respect to a "standard sediment" with 25% fraction < 63 μm and 10% OM content

Stations	Median grain size (phi)	Sorting coefficient (phi)	% < 63μm	% OM	Cr (mg/kg)		Cu (mg/kg)		Zn (mg/kg)		Pb (mg/kg)	
					M	C	M	C	M	C	M	C
LD1	2.8	1.4	36	25	335	275	290	172	25,790	18,469	2,972	2,678
LD2	0.3	1.4	23	25	101	105	123	116	247	271	216	273
LD3	3	2.1	8.7	6	550	816	139	174	230	371	487	739
LD4	0.5	2	6.9	7.2	65	101	103	138	194	340	457	727
LD5	3.2	2.4	6.3	6	63	101	131	188	187	348	209	345
WD1	-0.5	3.3	4.2	4.2	73	125	130	200	284	583	315	543
WD2	2.8	2.6	36	3.8	335	275	290	172	25,790	18,469	2,972	2,678

Heavy metal contents in sediments

Marked differences of heavy metal contents within the sediment were found across both rivers and most stations for the three monitoring occasions M1-M3 (Figure 3.3.). At LD2 and LD4, the maximum Cr content of 740 mg Cr/kg at LD4, exceeded the TEC and PEC guideline limits on all three occasions, and especially during M3 (Figure 3.3.a). Although the lowest sediment Cr contents were found at the two stations in the Worka River, the content was still close to the TEC guideline. Compared with the other Leyole stations, Cu contents (Figure 3.3.b) were notably higher downstream of the steel factory at LD2, by at least a factor of 2 and 8, especially for the M1 and M3 occasions. Here, the Cu contents exceeded the PEC in all monitoring occasions and, also downstream of the tannery factory (LD4) for two occasions (i.e. M2 and M3). The lowest Cu contents were observed for the Worka river stations, exceeding the TEC but not the PEC guidelines. As expected, highest Zn contents (Figure 3.3.c) were observed downstream of the steel factory (LD2) for M1-3, in all cases exceeding TEC and PEC; the same holds for Pb (Figure 3.3.d). TEC and PEC exceedances were often observed at other stations, in both rivers. Heavy metals sampled from the Leyole and Worka river sediments generally showed a decreasing trend with increasing grain size (Figure 3.4.), although with some considerable variation across sites. The Spearman rank correlation coefficients were related to

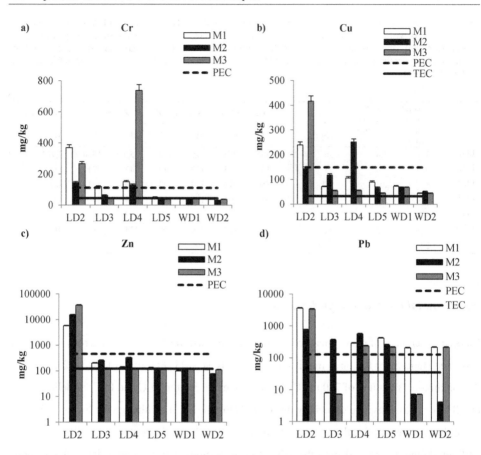

Figure 3.3. Heavy metals contents in mg/kg (mean ± standard deviations, n=3) on sediments samples collected from stations in Leyole and Worka rivers (Figure 3.1.c) in the three monitoring occasions M1-3. The different sediment quality guidelines are also presented here. (Note the logarithmic scale for c) and d).

all heavy metals, OM and grain size groups (Table 3.5.). Significant (p ≤ 0.05) positive correlations were found between Cr and Pb, and Cu and Zn, depicting their similar behaviour with respect to grain sizes and/or having a common source. Strongly significant (p ≤ 0.01) correlations were found between Pb and OM (Table 3.5.).

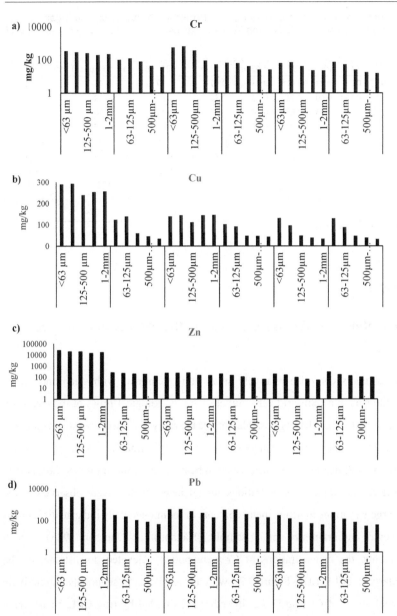

Figure 3.4. Heavy metals contents (a-d: Cr, Cu, Zn, Pb) in mg/kg (mean ± standard errors; n=3) in five grain sizes groups for sediment samples at the stations of the Leyole and Worka rivers (Figure 3.1.c), as average over three monitoring occasions M1-M3 (N.B. Cr, Zn and Pb concentrations are in logarithmic scale)

Table 3. 5. Spearman rank correlation coefficients (n= 18) for the heavy metal and organic matter contents, and sediment grain size fractions taken from six stations for three monitoring occasions c; cf. Figure 3.1.

	Cr	Cu	Zn	Pb	OM	< 63 μm	63-125 μm	125 – 500 μm	500 μm – 1 mm	1 – 2 mm
Cr		0.77	0.77	0.83*	0.91*	0.93*	0.89*	-0.17	-0.31	-0.71
Cu			0.83*	0.60	0.74	0.75	0.54	-0.61	-0.49	-0.71
Zn				0.83	0.85*	0.78	0.77	-0.12	-0.20	-0.49
Pb					0.97**	0.93**	0.94**	0.06	-0.37	-0.66
OM						0.99**	0.91*	-0.09	-0.41	-0.74
< 63μm							0.87*	-0.19	-0.49	-0.81
63-125 μm								0.12	-0.26	-0.60
125 – 500 μm									0.75	0.64
500μm - 1mm										0.89*
(1 - 2)mm										

*Significant at p ≤ 0.05; ** at p ≤ 0.01

3.4 Discussion

3.4.1 Heavy metals quality assessment in the Kombolcha Rivers and sediments

Kombolcha's catchments comprise relatively small sub-catchments with steep and gulley dissected flat landforms in a semi-arid agro-climate. The hydrological flows of the rivers are generally low and monitoring is hardly possible outside the rainy season. Our estimates provide preliminary seasonal estimates of heavy metals concentrations in water and sediment over two years, 2013 and 2014. While there were differences in heavy metal concentrations across many sites between the rivers, highest concentrations for all heavy metals were observed in the Leyole River. Large variations in heavy metal contents in sediments likely indicate differences in sediment structure and grain size among samples (Håkanson and Jansson, 1983). In both water and sediments, the concentrations of Cr were notably elevated downstream of the tannery, particularly in 2014 (LD4, Figure 3.2.) with accumulation in sediments. A relatively high partition coefficient of Cr (mg Cr/kg, in sediment) divided by (mg Cr/L, in water) = 740/2.66 = 280 L/kg; see Table 3.1.; Figure 3.3.) likely contributing to that (Allison and Allison, 2005). Elevated Cr downstream of a tannery is consistent with findings of Gebrekidan et al. (2009) elsewhere in Ethiopia and sub-Saharan countries (see Table 3. 6).

Significant temporal variations of Cu concentrations within and between sampling periods (Table 3.3.) likely shows infrequent discharges of Cu, which was highest in upstream parts of the Leyole River (LD2) compared with other stations (Figure 3.2.b and Figure 3.3.b) consistent with pollution from the effluent discharge from the steel processing factory (cf. Figure 3.1.c), as Cu salts are commonly used in heating and galvanizing steel (Jumbe and Nandini, 2009; Stigliani et al., 1993). Additionally, local open-pit, and unlined, landfills close to the river further upstream could be a sources of not only Cu but also other heavy metals (Table 3.1.), as solid wastes from the five factories are dumped there. Additional study is required to quantify pollution from these landfills.

Table 3.6. Maximum concentrations of selected heavy metals in river (mg/L) water and sediments (mg/kg) reported in selected sub-Sahara countries

sub-Sahara country	Cr		Cu		Zn		Pb		References
	Water	Sediment	Water	Sediment	Water	Sediment	Water	Sediment	
Egypt	0.06	185	0.05	333	0.12	743	0.02	95	Cu, Zn and Pb in water: (Abdel-Satar et al., 2017); Cr, Cu, Zn and Pb in sediment: (El-Bouraie et al., 2010)
Ghana	177	5.06	-	-	8526	17.87	42.7	9.36	(Afum and Owusu, 2016)
Nigeria	0.92	0.31	0.39	3.97	2.23	4.39	0.84	2.05	Heaavy metals in water: (Dan'azumi and Bichi, 2010); Heavy metals in sediment: (Sabo et al., 2013)
Tanzania	0.13	12.9	0.08	89.1	0.06	27.1	0.27	30.7	Heavy metals in water: (Kihampa, 2013) Heavy metals in sediments: (Kishe and Machiwa, 2003)
Uganda	0.02	103	6.3	78.3	3	351	3	90	(Fuhrimann et al., 2015)
Zimbabwe	2.48	16.1	0.23	38	0.50	100	1.02	41	For Cr: (Yabe et al., 2010) For Cu, Zn and Pb: (Nyamangara et al., 2008)
Standard Limits	0.05	25	2	18.7	4	124	0.01	30.2	Water: (WHO, 2011) Sediment: (USEPA, 1997b)

For both water and sediments of the Leyole River, the median Zn concentrations were highest in 2014 immediately downstream of the steel processing factory (Table 3.1., Figure 3.2.c); and this is likely due to both the larger scale production of Zn galvanizing of steel products in this factory during 2015 (personal communication, steel processing factory administration) and the legacy of effluent discharges from the zinc galvanizing process in previous years. Zn concentrations significantly varied both within and between sampling periods (Table 3.2.), suggesting periodic discharges of Zn.

The concentrations of Pb were highest in the downstream zone of the Leyole river, likely due to the tannery discharges (Zinabu et al., 2017a). Though the industrial sources in the upstream parts of the rivers were found low in Pb emissions (Zinabu et al., 2017a), the relatively high Pb accumulation in the riverbed sediments (Figure 3.4.) probably shows the incidence of Pb pollution from other source(s) such as vehicles that are washed beside the river (Karrari et al., 2012). The finding of often relatively low concentration of the heavy metals in water but high contents in the sediment (cf. Figure 3.2. and 3.3.) demonstrates a historical signal in sediments, while water results are more a "snap shot view" (Chapman, 1996). The tannery and steel processing factories have, respectively, being emitting Cr and Zn for several decades.

The runoff from the catchment areas of the rivers affect sediment grain size and, in turn, heavy metal contents found in the sediment. Highest heavy metal adsorption capacities can be expected for fine grained (< 63 μm) sediments, because of their larger specific surface area (Devesa-Rey et al., 2011). In both the Leyole and Work rivers, the correlations between the fraction < 63 μm and Pb, Cr and Zn contents were strong (Table 3.5.). However, the trend of decreasing heavy metal concentration with increasing grain sizes was not straightforward (Figure 3.4.). A clearer trend would probably show further differentiation within the <63 μm fraction (Devesa-Rey et al., 2011). The organic matter (OM), most likely from upstream agricultural lands and industrial wastes, may also have led to increased contents of Cr, "Cu, Zn and Pb on the sediments. The positive and strong correlation (except for Cu) found between the OM and heavy metal content supports this view (Table 3.5.). In a comparable river, Lin and Chen (1998) reported higher Pb concentration with increasing OM.

To account for the effect of OM and grain-size distributions in the content of heavy metals on sediments, it is important to normalize the variation based on these factors. Applying a normalization procedure as used in The Netherlands, comparing with a "standard sediment" with 25% fraction of < 63 μm and 10% OM content, yields results accounting for variation of heavy metals content owing to all existing ranges of grain-sizes and organic matter (Akele et al., 2016; Department of Soil Protection, 1994). This changes the estimated toxicity effect

(Table 3.4.), with the number of stations exceeding the PEC limit increasing from two to three for Cr; one to four for Cu; and one to two for Zn. Similarly, normalized Pb concentrations would increase at all stations, although the Pb concentrations at the stations already exceeded the PEC limit without normalization.

In general, the effects of industry on water quality in Kombolcha is common across Ethiopia and many parts of Africa. It is reminiscent of the situation in Europe and the U.S. before the proliferation of licensing and regulation that arose from the 1960s and 1970s in response to public concern and extreme and high profile pollution episodes (Hildebrand, 2002). To overcome the problems encountered in Ethiopia, it is important to understand the current situation of rivers and sediments and identify the needs and opportunities to build knowledge and capacity for better monitoring and management of rivers and other water bodies.

3.4.2 Efforts to improve monitoring of rivers' water quality in Ethiopia

With increasing uncontrolled waste discharges and anthropogenic inputs into sub-Saharan rivers, heavy metals pollution is of major concern (Peletz et al., 2018; Abdel-Satar et al., 2017). Several studies report that the heavy metals concentrations in rivers, pose high risks in many sub-Saharan countries (see Table 3.6.). As the dilution capacity of the rivers to pollutants is diminishing in the face of intensifying consumption of river water for irrigation, storage and global climate change (Abdel-Satar et al., 2017), the deterioration of the rivers and sediment quality is widespread (Zinabu et al., 2017a; Akele et al., 2016). A critical assessment of heavy metals and their impact on river water quality are thus pressing needs. This requires prioritizing water quality problems in regions in guiding water safety management, and ensuring environmental health through an effective policy framework that sets practicable monitoring and regulation of heavy metals in rivers and sediments.

In Ethiopia, two federal institutions deal with water quality monitoring. The first, EMEFCC, is mandated to establish national water quality criteria and pollution control policy (EEPA, 2010; FDRE, 2002a). The EMEFCC has developed national legally binding emission guidelines for eight categories of factories, and is authorized to extend technical guidance to water bureaus and regional environmental institutions to regulate the emission (Zinabu et al., 2017a). The guidelines are not specific for what they are designed to protect, although generally understood to protect the ecological wellbeing of water resources. While they are not specific for a particular water use, the application of international guidelines to protect water quality for the most common water uses in Ethiopia, for e.g., irrigation and livestock drinking water and protection of aquatic life, is recommended. The second federal institution, the Ministry of Water, Irrigation and Electricity (EMoWIE) and its institutions across regions, has a task of monitoring pollution of water resources and regulates in accordance with the Council of Ministers Ethiopia Water Resources Management Regulation (No. 115/2005) (EMoWR, 2004a). A policy of "polluter pays" is based on Water Resources Management Proclamation No.197/2000, but enforcement by regional and local water bureaus and environmental institutions is barely implemented because of a fundamental lack of capacity at the individual and institutional level (EMoWIE, 2016). Water quality standards of rivers, including those for heavy metals, merely follow the WHO guidelines for drinking water (EMoWR, 2004b).

Of Ethiopia's eight major river basins, only the Awash basin has any systematic monitoring (EMoWIE, 2016), due to its relatively higher economic significance compared with the other river basins, with a focus on water courses of economic importance and high industrial and agricultural activities. Both surface and ground waters samples e have been monitored over the last 15 years for standard hydro-chemical variables (total dissolved solids (TDS), dissolved oxygen (DO), Biochemical and Chemical oxygen demand (BOD, COD), electrical conductivity (EC) and pH) (personal communication, EMoWIE). River water samples are collected monthly from 17 stations and from six beverage producer effluents. The ground water, largely used for drinking water supply, is monitored at the abstraction points. Although large commercial farms (e.g. for the international flower market) in the Awash basin are reported to discharge a range of pollutants including heavy metals, pesticides and herbicides (Endale, 2011; Getu, 2009), there is no monitoring of nutrients or pesticides. Some sporadic, but extremely limited, monitoring of sediment, micro-pollutants and heavy metals have occurred, but with processing constrained by lack of suitable laboratory facilities. Data from all monitoring is stored at the federal level (in the EMoWIE) and used mainly for monitoring drinking water quality and industrial developmental projects enquiring for information. In addition to the effort of the EMoWIE, the EMEFCC established a water quality monitoring network in the Awash basin. The network is based on 36 sites along the (upper) Awash River. The EMoWIE deals with sampling and analysis of water quality and coordinates cross-sectorial activities. While both EMoWIE and EMEFCC are working for common purpose, lack of clear definition of roles and responsibilities between the institutions is preventing coordination to make better use of information.

With national policies that promote an industrial led economy and concurrent growing urbanization and intensive horticulture, there is an essential need for adequate monitoring for a suite of pollutants, including heavy metals, nutrients, organic matter and pharmaceutical compounds. While the Ethiopian government is attempting to mainstream the Sustainable Development Goals (SDGs) and allocate financial resources for effective coordination of their integration into the GTP-II (FDRE 2016), this cannot be effective without a supporting and relevant monitoring and management regime. The same applies to the ambition for fostering green industry development and encouraging socially responsible and environmentally safe sustainable manufacturing industries through building of industrial parks (FDRE, 2016).

In addition to the limited monitoring networks and regulation, low levels of financing for environmental research and monitoring has hindered policy-making (Awoke et al., 2016). This has hindered availability of reliable information on water quality and undermined the capacity to develop national water quality guidelines. While current economic development of Ethiopia is moving at a fast pace, the overall awareness and interest of environmental protection is still limited, and local application of water quality guidelines largely absent (Zinabu et al., 2017a; Damtie and Bayou, 2008). Drinking water quality following WHO guidelines, has some monitoring that focusses on physico-chemical and bacteriological components (Alemu et al., 2015). In most cases, due to limited capacity and lack of adequate laboratory instruments, heavy metals are left out in analyses (Zinabu et al., 2017a; EMoWIE, 2016). The WHO guidelines are not designed to assess ecosystem health, and many of the toxic substances included by the WHO have no legally mandated monitoring in Ethiopia (WHO, 2011).

Many sub-Saharan countries adopted water quality guidelines from developed countries and international guidelines often do not consider ongoing economic, social and technical needs (Ongley, 1993). Apart from some effluent regulations, few of the countries implement appropriate national water and sediment quality guidelines. Developing a new water quality paradigm based on these needs is necessary and the focus has to be on development of institutional capacity to formulate practicable guidelines and effective use of data. Ghana and Kenya use international guidelines to assist development of a national water standards. These countries used their own water quality data to derive national drinking water quality standards based on the WHO guidelines (Peletz et al., 2016). Another approach is to simply adopt another country's guidelines. In Nigeria, the Federal Environmental Protection Agency (FEPA), reviewed the water quality standards from Australia, Brazil, Canada, India, Tanzania, and the USA to derive Interim National Water Quality Guidelines and Standards, including those for aquatic life protection and irrigation and livestock watering (Enderlein et al., 1997). In Ethiopia, no national water quality standards are currently developed.

Based on our assessment of field reports of Ethiopia regions, limited financial sources hinder acquisitions of the necessary field and laboratory equipment, and human capacity is insufficient to deal with current water quality assessment needs of the EMoWIE. The lack of skilled professionals has hindered the use of information for better water quality management. Additionally, the absence of adequate monitoring infrastructures and aging of the existing ones are obstacles to establishing monitoring networks. It is clear that developing the capacity and financial needs of monitoring are basic factors to reverse the current trends of water quality degradation. This suggests the need for a policy focus on adaptable and affordable approaches for water quality monitoring.

3.4. 3 Monitoring in Transboundary Rivers and lakes of Ethiopia

Ethiopia has seven major transboundary rivers. In the face of the Ethiopia's commitment to the SDGs, it is timely to establish river and sediment quality monitoring networks in the rivers crossing neighbouring countries. Ethiopia is a member of the Nile Basin Initiatives (NBI), an intergovernmental partnership of ten Nile riparian countries, and endorsed designing of regional monitoring networks, including water quality (http://nileis.nilebasin.org/content/nile-basin-water-resources-atlas). Across the region, monitoring remains at an early stage and data exchange among stations and central data repositories is poor (Abdel-Satar et al., 2017; Nile Basin Initiative, 2016). The recent NBI development in designing a regional hydrometric system enhancing existing networks such as IGAD-HYCOS Program based on national needs and limitations, is an important action to improve the river basin monitoring systems (Nile Basin Initiative, 2016).

Although Ethiopia is a party to the agreement of the Nile basin initiative including monitoring of the Nile River for environmental sustainability (Nile Basin Initiative, 2016), the monitoring is still focused on meteorological measuring of daily rainfall and temperature, and hydrometric measuring of river or lake water levels. Monitoring of water quality, sediment transport in rivers, and groundwater are yet to start in most member countries (Nile Basin Initiative, 2016), while at the same time rivers are increasingly used for irrigation and energy generation (Abdel-Satar et al., 2017). This needs immediate attention, as the conditions can pose pollution risk and affect chemical process and ecological health downstream. Although some of seven major transboundary rivers of Ethiopia eventually drain into the Nile River and are hence included in the Nile Basin Initiative, three major rivers (the Omo, Shebelle, and Mereb rivers) flow across the boundaries with Kenya, Somalia and Sudan, respectively. There is no clear international collaboration for monitoring these rivers. Additionally, Ethiopia has twelve big lakes. Apart from the Lake Tana, which is source of the Nile River, most of the lakes are found in the rift valley of the country. Population pressure and rising economic interest places increased pressures on lake water (Francis and Lowe, 2015). Larissa et al. (2013) report high heavy metals concentrations both in the water and tissue of edible fish species found in the Koka and Awassa lakes, which are receiving industrial effluents.

3.5 Conclusion and the way forward

Chromium, Cu and Zn were generally found at toxic concentrations in both water and sediments of the Leyole and Worka rivers of Kombolcha. Cu and Zn concentration varied between the sampling periods while Cr consistently varied between the upstream and downstream of the Leyole River. The old tannery and steel processing industries are likely sources for enhanced Cr and Zn in the Leyole River and sediments. More study is required to validate emission of heavy metals from local landfills. This study has demonstrated that management of heavy metal pollution is a serious issue in Kombolcha, a city that illustrates the problems across Ethiopia and the region more generally. To ensure effective monitoring of rivers and sediments and provision of monitoring information, commitments from jurisdictions is needed in developing institutional capacity and implementing adaptable, including low-cost, monitoring techniques.

There is a clear need for compliance to water protection requirements, and feasible guidelines responsive to the environmental problems at hand. This could be either by reviewing applicability of international or neighbouring countries policies and guidelines or, as done in Ghana and Kenya, building a monitoring network that provides baseline data to inform national policy. In Nigeria, the reviewing process included consultation with a range of interested parties, including those from the private sectors, higher education institutions, non-governmental organization and the interested public (Enderlein et al., 1997).

Low-cost techniques and, advocacy for public and private partnership can help overcome limited governmental institutional structures, lack of adequate instruments, and deficiencies in necessary skills in river monitoring. Growing use of information technology using smart phones provides an example of new opportunities that can both transmit monitoring data and raise local awareness of the importance of water quality monitoring. This can readily link to GIS and remotes sensing techniques to support modelling of river water quality and monitoring of rivers in relatively large basin areas (Dube et al., 2015; Ritchie et al., 2003). Coupled with modelling of pollutant emissions from point and non-points sources provides the potential for improved risk assessment of pollutants in rivers (Zinabu et al., 2017b; Brack, 2015; Daughton, 2014).

Cost effective monitoring networks as described above can strengthen both local institutional capacity, and engagement of citizens in science, even in remote and rural areas (Katsriku et al., 2015; Danielsen et al., 2011). Simple monitoring techniques such as mini stream assessment scoring system (www.groundtruth.co.za/projects/minisass.html) applied in South Africa is an example of implementing adaptive managements at local scales (Aceves-Bueno et al., 2015). An example of a Citizen Science project implemented in countries with varied social and economic structures, including Kenya and Zambia is the EU funded Ground Truth project (http://gt20.eu/). This can also apply in remote areas of Ethiopia where infrastructure is poor and data capturing challenging (Katsriku et al., 2015). In Senegal, a mobile phone application implemented to facilitate water quality data collection within the national public health agency has been an effective intervention (Kumpel et al., 2015).

Chapter 4

Evaluating the effect of diffuse and point source nutrient transfers on water quality in the Kombolcha River Basin, an industrializing Ethiopian catchment

Publication based on this chapter:

Eskinder Zinabu, Peter Kelderman, Johannes Kwast, Kenneth Irvine. 2018. Evaluating the effect of diffuse and point source nutrient transfers on water quality in the Kombolcha River Basin, an industrializing Ethiopian catchment. Land Degradation and Development 29:3366-3378.

Abstract

Many catchments in sub-Saharan Africa are subject to multiple pressures, and addressing only point sources from industry does not resolve more widespread diffuse pollution from sediment and nutrient loads. This paper reports on a preliminary study of nutrient transfers into rivers in two catchments in the industrializing city of Kombolcha, North Central Ethiopia. Sampling of rivers and industrial effluents was done over two sampling periods in the wet season of 2013 and 2014. Catchments boundaries and land use map were generated from remote sensing and ground data. Higher total nitrogen (TN) concentrations were found from sub-catchments with largest agricultural land use, whereas highest total phosphorus (TP) was associated with sub-catchments with hilly landscapes and forest lands. Emissions from brewery and meat processing were rich in nutrients (median TN: 21–44 mg L^{-1}; TP: 20– 58 mg L^{-1}) but contributed on average only 10% (range 4–80%) of the TN and 13% (range 3– 25%) of the TP loads. Nutrient concentrations in the rivers exceeded environmental quality standards for aquatic life protection, irrigation, and livestock water supply. In Ethiopia, more than 85% of farmers operate on less than 2 ha of land, with concomitant pressure for more intensive farming. Land is exclusively owned by the State, reducing a sense of land stewardship. As the City of Kombolcha moves to agricultural intensification and increased industrialization, attention is needed to fill gaps in monitoring of nutrient pollution in rivers and use information to reconcile development with land use and its degradation.

4.1 Introduction

The impact of nitrogen and phosphorus on surface water quality is an increasing problem in developing countries (Mustapha and Aris, 2012). Monitoring and reporting on nutrient emissions are often absent, insufficiently reported, or of uncertain quality. In urban and peri-urban areas, nutrient loads come from both point increasing problem in developing countries (Moges et al., 2016); and diffuse sources. In many parts of sub-Saharan Africa, neither are monitored effectively (Dybas, 2005).

In many African countries, increasing and more intensive cultivation of land, driven by rising populations, leads to land degradation, notably soil erosion, mineral depletion, and altered patterns of surface run-off and infiltration. These pressures can reduce nutrient retention in soils and increase nutrients loads into surface waters (Delkash et al., 2018). Intensive cropping and increased use of fertilizer and animal manures increase nutrient loads into surface waters (Duncan, 2014; Gasparini et al., 2010). Soil erosion enhances transfer of particulate phosphorus and nitrogen from land to water (Soranno et al., 2015). In urban and peri-urban areas, discharges from domestic, industrial, and waste water treatment provide typical point sources of nutrients to surface water (Tasdighi et al., 2017; Mohammed, 2003).

In sub-Saharan countries, the vulnerability of surface water to nutrient pollution is often either not well considered or understood (Nyenje et al., 2010). Ethiopia exemplifies many of these challenges. Government policy promotes more intensive agriculture together with a drive for industrialization. Existing government policies to prevent industrial pollution are often poorly implemented (Gebeyehu, 2013). Agriculture increasingly encroaches marginal lands, and cultivation of hillslopes promotes land degradation (Gashaw et al., 2014).

This study reports preliminary research on nutrient transfers in two river catchments in the industrializing conurbation of Kombolcha City in North Central Ethiopia. Kombolcha is topographically varied with a rural upland landscape and lowland urban areas that are both prone to erosion (see section 1.3.1). The Leyole River catchment, in the upper part of this study area, consists of the Tebissa, Ambo, and Derekwonz sub-catchments (Figure 4.1.). In the middle section of the Leyole River, four factories discharge effluent into the river, and new factories are being planned. The Worka River receives effluent from a brewery. Wastes from urban household latrines are exported to a site 5 km away from the town, but human open defecation, also contributing nutrient loads to the rivers, is common. The downstream sections of both rivers are used for irrigation and watering livestock.

Kombolcha typifies a situation common throughout sub-Saharan Africa, providing a useful case-study for assessing combined effects of increasingly intensive land use and industrial development. Formal monitoring of water quality of the rivers by local or regional authorities does not occur. This study assesses multiple pressures on water quality from a combination of land use intensification, poorly regulated industries, and human inputs from open defecation. We (a) conducted a preliminary water quality sampling study, during two successive rainy seasons, in the Leyole and Worka rivers and (b) using simple models estimated potential contributions of diffuse nutrient loading from land use and human open defecation.

4.2 Materials and Methods

4.2.1 Sampling locations, times and collection

The main rains in Kombolcha, Ethiopia (Figure 4.1.) fall from June to September (690 mm in 2013; 711 mm in 2014). These follow an earlier rainy season, typically from February to April (109 mm in 2013; 129 mm in 2014), but which has become erratic and of low intensity in recent years, leading to recurrent droughts with high annual potential evapotranspiration (2014: 3,046 mm; Kombolcha Meteorological Branch Directorate, 2015). These factors guided the timing of sampling, done between June 15 and September 30, both in 2013 and 2014. Owing to financial and logistic constraints, sampling was done at approximately 15-day intervals. Sampling of nutrients from factory emissions was done concurrently with that of the rivers. The factories show typical year-round emissions (Zinabu et al., 2017a). River water samples were collected at the outlet of four sub-catchments and two catchments (Figure 4.1.). The samples were collected from approximately 30 cm depth at 1/4, 1/2, and 3/4 of the width of the river. A composite sample was then prepared from equal volume proportions and stored in a 100-ml polyethylene container. On each sampling occasion, composite samples were taken from a series of grab samples from factory effluent discharge pipes using a 100-ml polyethylene container. Depending on the way the effluents were discharged, the composite grab samples were taken at equally spaced time interval or equal volume. Using information on the timing of effluent discharges acquired from the factories' administrations, three grab samples were collected at the beginning, halfway through, and at the end of the discharge periods from factories with intermittent batch discharge of effluents (steel processing, meat processing, and tannery). For factories with continuous effluent discharges (brewery and textile), eight grab samples were taken every 3 hr in a 24-hr period of the sampling day, and samples were mixed in equal batches.

4.2.2 Measuring nutrients on site and in laboratory

After filtration of the samples over Whatman glass microfiber filters (pore size about 1.3 μm), a Hach DR 890 colorimeter field kit was used for on-site analysis of nitrate (NO_3-N), ammonium (NH_4-N) and soluble reactive phosphorus (PO_4-P). The Hach apparatus had earlier been tested at the IHE Delft Institute using standards of known concentrations of samples and

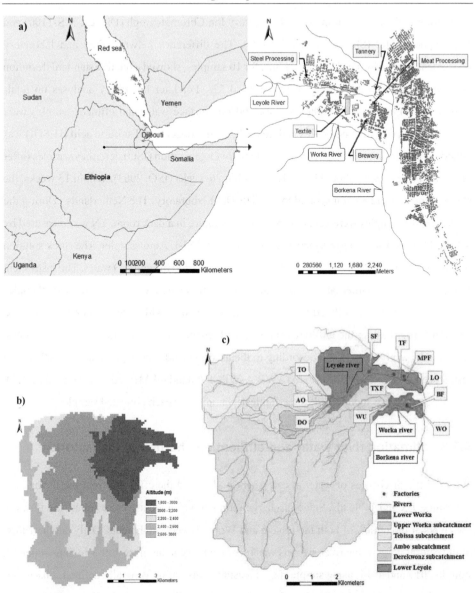

Figure 4.1. (a)Location of the Kombolcha city administration in north-central Ethiopia, east Africa in the left-side and including the study area in the right-side; (b) surface elevation of lands in the Kombolcha city administration; and (c) locations of the monitoring stations (i.e., at catchment outlets of the Leyole River and Worka River and factories effluent discharge pipes [AO: Ambo sub-catchment main stream outlet; DO: Derekwonz sub-catchment main stream outlet; LO: Leyole river downstream; TO: Tebissa sub-catchment main stream outlet; WO: Worka river downstream; WU: Upper Worka sub-catchment outlet]), sub-catchments, and rivers in the Leyole and Worka river catchments; SF: Steel processing factory; TF: textile factory; TF; Tannery factory: MPF: meat processing factory; BF: Brewery factory

comparing with analysis from both a Laboratory Ion Chromatograph (Dionex, ICS-1100) and Spectrophotometer (Shimadzu, UV-250IPC). The difference between field and laboratory instrument was < 5%. The Hach apparatus, for 10 samples, showed a relative standard deviation (RSD) among replicate samples between 1and 5%. For later laboratory analyses on total-nitrogen (TN) and total-phosphorus (TP), 100 mL surface water and effluent samples were transferred to polyethylene bottles and, to each, 1 mL concentrated sulfuric acid (H_2SO_4) was added for preservation. The International Standard Organization (ISO) describes samples (after preservation in H_2SO_4) should be analysed within 26 weeks (ISO, 2003). Within 15 weeks, the above samples were air-transported to the IHE Delft laboratory, The Netherlands. During the 15 weeks, the samples were kept at sub-zero temperature in a dark room. TN was measured by a TOC-L (type: L-CPN) Shimadzu apparatus with an ASI-L Auto sampler. The stock solution for calibration was 1000 mg N L-1 and 7.22 g KNO_3 in 1000 mL milli-Q water. The instrument was checked with an internal reference, and the detection range was 0.5 – 20 mg L-1. Samples with > 20 mg L-1 were diluted with Milli-Q water down to within detection range. TP was measured after boiling the unfiltered samples with concentrated sulfuric acid for 30 minutes, followed by analysis as PO_4-P according to the ascorbic acid spectrophotometric method. All test procedures followed the APHA-AWWA-WPCF "Standard Methods" (Rice et al., 2012). Quality control was maintained using duplicate samples for each collected sample.

4.2.3 Estimating river water and effluent discharges of the factories

We used the discharges of water at catchment outlets and the effluents from the factories to determine TN and TP loads. Stage-discharge rating curves were used to estimate the flows of the rivers at the outlets of both the Leyole and Worka Rivers catchments. Daily depth of flow at the central point of the river stations was recorded twice a day from July 1 to September 30, both in 2013 and 2014, using simple stage measurements within vertically defined subsections across the river. At each station used to estimate flows, the river channel cross section was divided into multiple vertical subsections and the velocity of flow at each subsection determined using a current meter (Price-Type AA, vertical axis cup) and pigmy-current metre, for medium-high and low flows, respectively. For each sampling occasion, 12 discharge estimates were made by multiplying the velocity of the water flow and the cross-sectional area of the river. The discharge in each subsection was determined by multiplying the subsection area by the estimated velocity, and total discharge was computed by summing discharges for each subsection. Stage-discharge rating curves were prepared following the US Geological Survey (USGS) technical assistance for Hydrology Project TWI 3-A1 (Kennedy, 1984). The coefficients of the rating curves were estimated using least mean square methods, and estimates of flows were derived from these (Das, 2014). Similarly, the industrial effluent discharges were measured using a volumetric method, because this gives most accurate results for very small flows as at the outfall of a pipe (Hamilton, 2008). The time to fill a fixed volume, within a 40-L container, was first estimated for each factory's effluent discharge pipe. Flow rates were then calculated by dividing the volume by the time to fill up the container.

4.2.4 Spatial analysis

Sub-catchment delineation and land use mapping

The sub-catchments and catchments boundaries were derived from the Shuttle Radar Topographic Mission 1 with the arc - second global digital elevation model (USGS, 2006). Three sub-catchments were delineated in the Leyole and one in the Worka River (Figure 4.1.). The digital elevation model was re - projected to Universal Transverse Mercator zone 38N, Datum - Adindan. The BASINS 4.1 (Better Assessment Science Integrating Point and Nonpoint Sources), a multipurpose GIS freeware that integrates environmental data analysis tools and modelling systems, was used for catchment delineation to define land areas contributing to flows (USEPA, 2015). The stream network for the catchment was downloaded from USGS HydroSHEDS (Lehner et al., 2008). The points file containing sampling points of the catchment and the sub-catchment outlets were created using Global Positioning System readings at the catchment and sub-catchment outlets. These points were used to delineate the sub-catchment and catchment boundaries (Figure 4.1.). Land use and land cover map was derived using remote sensing analysis using MultiSpec 3.4 (Landgrebe, 1998). Field visits were undertaken in the study area to collect training and test areas for remote sensing image classification using a handheld GPS. The land use was categorized into seven groups using the USGS classification system (Anderson, 1976): bare lands; crops; forest; grazing land; residential areas; industrial areas; and water bodies. Based on the crop calendar of the study area, the least cloud cover Landsat 8 image (with surface reflectance bands product, band 7, level IT data type and grid cell size reflective of 30 m) of October 2014 was chosen. This helped to separate the crop lands from grass and bare lands. The image was terrain-corrected and filtered to account for some sensor variations caused by hills and valleys (NASA, 2014). Pixel based classification techniques were applied to map land uses using MultiSpec 3.4 (Landgrebe, 1998), a freeware image data analysis system by Purdue Research Foundation (Ndimele et al.). A maximum likelihood classification was applied using training areas collected from the study areas. The accuracy was assessed as outlined by Congalton and Green (2008), using the independent test areas, resulting in an overall accuracy of 89%, and a Kappa of 86%. Finally, the land-use map was polygonised in ArcMap 10 (ESRI, 2011). The area covered by each land-use class and the percentage of the total area covered were calculated in BASINS.

4.2.5 Statistical techniques

The sampling data sets were processed in Microsoft Excel and R statistical packages (R Core Team, 2015). Median concentrations were used for all stations' results (Ilijevic et al., 2015), and Grzetic, 2015). Multiple comparisons were used to test significant differences in nutrient concentrations among upstream and downstream stations in the catchments. To reduce effect of outliers and account for the relatively small size of samples used in each sampling programme (n = 8), the non - parametric Kruskal–Wallis test was chosen to compare significance of difference of nutrients among stations, followed by Dunn's post hoc test (non - parametric and two - group comparison) to determine significance of differences between stations (Niño de Guzmán et al., 2012).

4.2.6 Estimation of nutrient loads at the catchment outlets

Estimates of nutrient loadings at the outlets of the Leyole river and Worka river catchments (Figure 4.1.) were computed using Flux 32, a Windows - based interactive freeware developed by US Army Corps of Engineers in collaboration with the Minnesota Pollution Control Agency (Walker, 1999). Because the flow and nutrient concentrations were time series, a regression approach ('Method 6' in the Flux 32 software) estimated the loadings. A stratification scheme, based on coefficient of variation and residual errors from the regression, improved estimates of loading. For loading estimates having CV (coefficient of variation) > 0.3 and large residual errors (slope > 0.5; slope significance <0.5), a stratification scheme was used based on season (days) to minimize CV and residual errors. For cases that "Method 6" calculated CV> 0.3 after stratifications, weighted averages (i.e. "Method 2" in Flux 32 software) were used to estimate the loadings, as this method performs best for cases in which the natural flows and concentrations of a pollutant in a river are affected by point source discharges (Walker, 1999; Walker, 1987). It is also well suited for cases of many flows, but with few concentration data (Quilbé et al., 2006; Preston et al., 1989). The loadings by the factory effluents are not expected to vary much with flows. Therefore, a "direct load averaging" method was used to estimate the nutrients loadings in the effluents, taking the median of the loadings estimated for each sampling dates within the respective campaign (Walker, 1987).

4.2.7 Estimation of TN and TP loads from open defecation

An extensive literature review by Gumbo (2005) presents ranges of TN and TP contents in human excreta from developed countries in Europe and North America. As TN and TP in human excreta are affected by nutritional intake (Bouwman et al., 2005; Gumbo, 2005; Jönsson et al., 2005), we assumed half of the average TN and TP content in the excreta reported by Gumbo (2005) as reasonable values for Ethiopia (Table 4.4.). These values are comparable with data from Kenya, Tanzania, Uganda, and West Africa (Kelderman et al., 2009; Scheren et al., 2000). Assuming a 20% nitrogen loss as ammonia (Kimura et al., 2004), we estimated daily per capita "human loads" of TN and TP (kg day^{-1}; Table 5.3.). For human physiological reasons, assumed nitrogen contents in excreta are higher than phosphorus (Table 4.4.). As no previous studies that estimated daily export of "human loads" were found for comparable catchments of Africa, we applied transfer function estimates of 16% TN and 3% TP as generalized coefficients used by Johnes (1996) for hilly catchments in the United Kingdom.

4.2.8 Source apportionments

Nutrient emissions from land use, point sources, natural base flows of streams, and atmospheric deposition to surface water were used to apportion nutrient loads between point or diffuse sources. The sewerage wastes collected and stored about 5 km away from the lower catchment boundary of the study area are assumed not to contribute loadings to the Leyole and Worka River. The five factory discharges were considered as point sources. Surface flows comprised background nutrient transfers (i.e., natural land including forest and bare land), transfers from agriculture (e.g., crop and grazing lands), transfers from scattered villages' areas, and some small atmospheric deposition into the steams in the catchment. We estimated diffuse sources apportionment by subtracting the total loads measured at the Leyole and Worka catchments' outlets from the point source loads of the factories.

4.3 Results

4.3.1 Nutrient discharges from factory effluents

The estimated daily average factory effluent discharges for the five factories and daily nutrient concentrations and loadings for the two sampling studies, C1 (2013) and C2 (2014), are shown in Table 4.1. Median TN concentrations in the meat processing factory and brewery effluents had comparable values (C1: 36 and 44 $mg\,N \cdot L^{-1}$; C2: 21 and 36 $mg\,N \cdot L^{-1}$, respectively), with much lower TN concentrations for the other three factories (Table 4.1.).

Table 4.1. Estimates of the factories' effluent nutrient concentrations, discharges, and loadings into the Leyole and Worka Rivers during the first (C1) and second (C2) sampling programme, with n = 8 each, from June to September 2013 and 2014, respectively

Factory			Brewery		Meat processing		Textile		Tannery		Steel Processing	
Campaign			C1	C2	C1	C2	C1	C2	C1	C2	C1	C2
Effluent	Mean discharge	$L\,s^{-1}$	8.2	21	11	8.8	15.4	16.5	6.8	8.4	1.7	2.2
(NH₄ + NH₃)-N	Median	$mg\,L^{-1}$	0.4	1.3	1.5	1.4	0.2	0.01	0.04	0.18	0.14	0.1
	Minimum		<0.01	0.1	0.5	0.6	0.06	<0.01	<0.01	0.08	0.05	0.01
	Maximum		1.2	1.1	2.3	1.7	0.25	0.01	0.12	0.22	0.22	0.08
	Loadings	$kg\,day^{-1}$	0.4	1.1	1.1	1.1	2.57	0.1	0.04	0.14	0.1	0.08
NO₃-N	Median	$mg\,L^{-1}$	2.2	1.5	1.4	0.3	0.16	0.01	0.22	0.04	0.13	0.12
	Minimum		1.1	0.1	<0.01	0.01	0.01	<0.01	0.11	<0.01	<0.01	0.01
	Maximum		11	4.0	3.1	1.2	0.34	0.03	1.08	0.16	0.29	0.31
	Loadings	$kg\,day^{-1}$	4.0	4.0	2.1	0.4	0.23	0.03	0.39	0.05	0.20	0.31
TN	Median	$mg\,L^{-1}$	44	36	36	21	4	2.8	4.30	2.7	3.4	2.8
	Minimum		15	10	22	4	2.2	0.8	2.20	1.0	2.1	1.8
	Maximum		89	49	99	32	6.2	9.1	5.40	9.0	5.20	11.7
	Loadings	$kg\,day^{-1}$	33	88	36	15	5.3	3.9	2.5	1.9	0.5	0.5
PO₄-P	Median	$mg\,L^{-1}$	4.1	2	5.1	2.1	0.09	0.4	0.09	0.36	0.04	0.34
	Minimum		4.1	1	5.1	1.2	0.09	0.4	0.09	0.2	0.04	0.19
	Maximum		5.2	2	6.1	2.1	0.11	0.69	0.1	0.36	0.05	0.34
	Loadings	$kg\,day^{-1}$	3.1	3	4.0	1.3	0.07	0.43	0.07	0.22	0.03	0.21
TP	Median	$mg\,L^{-1}$	20	58	32	33	0.4	11.2	0.55	5.7	0.2	5.4
	Minimum		4.1	5.1	5.1	4.8	0.11	1.3	0.23	0.1	0.02	0.03
	Maximum		27	91	61	61	2.61	20.6	0.89	8.7	0.84	17.6
	Loadings	$kg\,day^{-1}$	35	28	37	28	0.5	15.9	0.32	4.1	0.04	1.1

Note. For the loading, the 'direct averaging loading method' was used. TN: total nitrogen; TP: total phosphorus.

Large differences were found between concentrations of TN and the sum of $(NH_4 + NH_3)-N$ and $NO_3 -N$ in the two factories' effluents, indicating dominance of other forms of nitrogen such as organic nitrogen amine forms. As for TN, median TP concentrations in the meat processing factory and brewery effluents were comparable (C1: 32 and 20 mg P·L−1; C2: 33 and 58 mg P·L−1, respectively; Table 4.1.). Here, however, the textile factory had an appreciable median TP contribution (11.2 mg P·L−1) during the C2 study. The TP concentrations were much higher than those of PO_4-P, showing the dominance of other forms of phosphorus in the effluents, most likely as particulate phosphorus (see later discussion).

4.3.2 Land use distribution in the catchments

Cropland was the largest land use in the four sub-catchments (Figure 4.2, Table 4.2.). Next highest was barren land coverage (21% to 25%), largely in the upper sub-catchments where overgrazing and deforestation on the hillsides were evident. Forest land in the Upper Worka and grazing land in the Tebissa sub-catchments comprised 8.4% and 1.6% of land use, respectively.

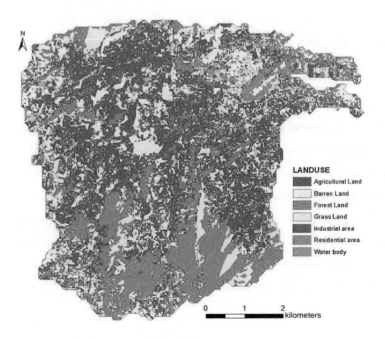

Figure 4.2. Map of land use in the Leyole and Worka river catchments

Residential and industrial areas were located mainly in the lower part of the river catchments. Waterbodies, mostly swampy areas and reservoirs for industries, were found only in the lower part of the catchments (Figure 4.2.).

Table 4.2. Land use and its percentage coverage of the Tebissa, Ambo, and Derekwonz sub-catchments and the Leyole and Worka river catchments

Land use	Sub-catchment								Catchment			
	Tebissa		Ambo		Derekwonz		Upper Worka		Leyole		Worka	
	Area	%	Area	%	Area	%	Area	%	Area	%	Area	%
Barren land	1.8	21.6	0.8	25	0.3	21	5.8	19.1	4.5	25.8	6.1	19.8
Crop land	3.6	44	1.7	55	0.7	57	11	36.4	6.8	39.2	11.2	36
Forest land	1.1	13	0.3	8	0.2	12	8.4	27.6	1.7	10	8.4	27
Grazing land	1.6	20	0.4	11.7	0.1	9.6	4.3	14.4	2.9	16.7	4.5	14.6
Village areas	0.1	1.4	<0.1	0.3	<0.1	0.4	0.6	2	0.2	1.1	<0.1	0.5
Industrial area	-	-	-	-	-	-	-	-	1.1	6.	0.7	2.4
Water areas	-	-	-	-	-	-	<0.1	0.5	<0.1	0.2	<0.1	0.2
Total area	8.2		3.2		1.2		29.7		17.3		31	

Note. Areas are in km^2.

4.3.3 Nutrient concentrations at catchment and sub - catchment outlets

The $(NH_4 +NH_3)$–N, NO_3–N, and TN concentrations showed large temporal variations, indicating both variations in water discharges and nutrient loadings (Figure 4.3.). $(NH_4 +NH_3)$–N concentrations (Figure 4.3.a) fluctuated from a minimum of 0.02 mg L−1 in the upper sub - catchments to a maximum of >2 mg L−1 at the outlets of both the Leyole and Worka catchments. Compared with (NH_4+NH_3)–N, variations of NO_3–N concentrations were smaller, with highest value in the Tebissa sub - catchment outlet (median value 2.8; maximum 4.5 mg L−1; Figure 4. 3.b). Intermediate concentrations of around 1mg NO_3–N·L−1 were observed at the other stations. In contrast to (NH_4+NH_3)–N, no marked differences in nitrate were observed between upstream and downstream in the Leyole and Worka Rivers. Similarly, TN differences were low among the four sub-catchment and between the catchments outlets (Figure 4.3.c).

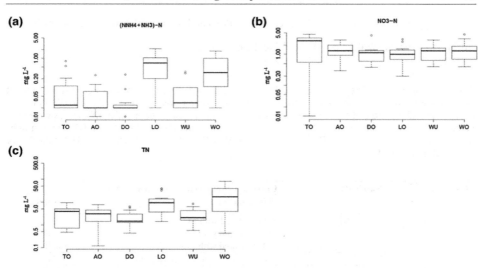

Figure 4. 3. Box plots of (a) (NH₄+NH₃)–N, (b) NO₃–N, and (c) total nitrogen (TN) concentrations (mg L−1) in the water samples (n = 16) collected over the two monitoring periods at the sub-catchments and catchment outlets of the Leyole River and Worka River. Each box encloses the interquartile range; the whisker bars represent minimum and maximum concentrations at a station. The median value is the second quartile inside the boxes (bold horizontal line within the boxes). Outliers (if present) are depicted as "o" above and below the whisker lines. AO: Ambo sub-catchment main stream outlet; DO: Derekwonz sub-catchment main stream outlet; LO: Leyole river downstream; TO: Tebissa sub-catchment main stream outlet; WO: Worka river downstream; WU: Upper Worka sub-catchment outlet

Median TN concentrations at the outlets of the Leyole and Worka river catchments were 9.3 and 17.3 mg L−1, respectively, with a maximum of 81 mg L−1 at the Worka River catchment outlet. Additionally, at these outlets, median TN concentrations were higher compared with the sum of (NH₄ +NH₃)–N and NO₃–N. We observed frequent peak concentrations (depicted as upper outliers above the Whisker line of the box plots, cf. Figure 4.3.a, b) of the (NH₄ +NH₃)–N and NO₃–N concentrations at the Derekwonz sub-catchment outlet. In line with above, the median (NH₄ +NH₃)–N concentrations was significantly higher in the outlet of Leyole river compared with the upstream sites (Tebissa, Ambo, and Derekwonz sub - catchments; Table 4.3.).

Median TN was significantly different (Kruskal–Wallis; p<0.005) only between the Derekwonz sub - catchment and Leyole catchment outlet. In contrast, median NO₃–N concentrations at the outlets of the upper three sub - catchments of Tebissa, Ambo, and Derekwonz and the Leyole catchment outlets were not statistically different from each other (Kruskal–Wallis; p> 0.01). For the Worka river catchment, only median NO₃–N concentrations at the Upper Worka sub - catchment and Worka catchment outlet were not significantly different (Kruskal–Wallis; $p > 0.05$; Table 4.3).

Table 4.3. Significant difference test Kruskal–Wallis (by rank) and Dunn's post hoc test for nutrient concentrations between upstream and downstream stations within the Leyole and Worka river catchments (cf. Figure 5. 1.; n =8); AO: Ambo sub-catchment main stream outlet; DO: Derekwonz sub-catchment main stream outlet; LO: Leyole River downstream; TN: total nitrogen; TO: Tebissa sub-catchment main stream outlet; TP: total phosphorus; WO: Worka river downstream; WU: Upper Worka sub-catchment outlet.

| Catchment | Nutrients | Upstream compared with downstream | |
		Paired station	p-values
	$(NH_4 + NH_3)$-N	TO compared with LO	<0.01
		AO compared with LO	<0.001
		DO compared with LO	<0.001
	NO_3-N	(TO, AO, DO) compared with LO	> 0.05
Leyole river	TN	DO compared with LO	<0.05
		(TO, AO) compared with LO	> 0.05
	PO_4-P	TO compared with LO	<0.001
		AO compared with LO	<0.05
		DO compared with LO	<0.01
	TP	(TO, AO, DO) compared with. LO	> 0.05
	$(NH_4 + NH_3)$-N	WU compared with WO	<0.05
	NO_3-N	WU compared with WO	> 0.05
Worka river	TN	WU compared with WO	<0.05
	PO_4-P	WU compared with WO	<0.001
	TP	WU compared with WO	> 0.05

Similar to nitrogen, peak concentrations (shown as outliers above the Whisker line, Figure 4.4. of PO_4–P and TP were observed at the Derekwonz outlet. The median PO_4–P concentrations were markedly higher at the outlet of the Leyole River catchment (1.4 mg L^{-1}) compared with the upper three sub - catchment outlets (0.27–0.56 mg L^{-1}; Figure 4.4.a). The same holds when comparing the upper with the lower Worka river catchment (median concentrations: 0.35 and 2.2 mg PO_4–P·L^{-1} at Upper Worka sub - catchment outlet and Worka River catchment outlet, respectively). For the Worka catchment outlet, a maximum of 5.9 mg PO_4–P·L-1 was observed. In contrast to PO_4–P, no marked differences were observed among the median TP concentrations at the three Leyole upper catchment stations and outlet Leyole catchment outlet (Figure 4.4.b). Generally, PO_4–P was only a small fraction of TP. For the Worka catchment, a marked difference in median TP concentration was observed between the upstream (Upper Worka catchment outlet: 3.8 mg P·L-1) and downstream (Worka catchment outlet: 22.6 mg P·L^{-1}), as well as maximum TP concentrations.

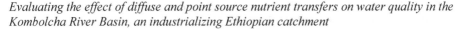

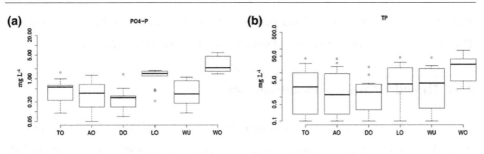

Figure 4.4. Box plots of (a) PO₄–P and (b) total phosphorus (TP) concentrations of surface water samples over the two monitoring periods (n = 16) taken at sub-catchments' outlets and downstream of the Leyole and Worka rivers. AO: Ambo sub-catchment main stream outlet; DO: Derekwonz sub-catchment main stream outlet; LO: Leyole river downstream; TO: Tebissa sub-catchment main stream outlet; WO: Worka River downstream; WU: Upper Worka sub-catchment outlet

The median PO₄–P concentrations at the three sub-catchment stations were significantly different (Kruskal–Wallis; p<0.001) from the Leyole catchment outlet (Table 4.3.). In contrast, there was no significant difference (Kruskal–Wallis; p > 0.05) in median TP concentrations between the Leyole's sub-catchments and catchment outlet, or between the upstream and downstream outlet of the Worka River (Table 4.3.).

4.3.4 TN and TP loads from open defecation

With a larger population, the TP and TN contributions estimated from human excreta are expected to be higher in the Tebissa and Upper Worka sub-catchments than the Derekwonz and Ambo sub-catchments (Table 4.4.).

Table 4.4. Estimated exported loadings (kg day⁻¹) of TN and TP in human urine and faeces in the Kombolcha sub-catchments following Gumbo (2005)ᵃ

Sub-catchment	Area km²	Inhabitants	Assumed compositions				Urine contribution		Feces contribution		Urine + feces		Exported into streams	
			Urine		Feces		TN	TP	TN	TP	TNᵇ	TP	TN	TP
			TN	TP	TN	TP								
			g (person. day)⁻¹							kg day⁻¹				
Upper Worka	29.7	20196					111	10	20	5	131	15	21	0.5
Tebissa	8.2	9783					54	5	10	2	64	7	10	0.2
Ambo	3.2	3806	5.5	0.5	1	0.25	21	2	4	1	25	3	4	0.1
Derekwonz	1.23	1473					8	0.7	1.5	0.4	9.5	1	1.5	0.03

Note. TN: total nitrogen; TP: total phosphorus.
ᵃContribution of TN in urine: 11 g·person⁻¹ day⁻¹; contribution of TP in urine: 1 g·person⁻¹ day⁻¹; contribution of TN in feces is 2 g·person⁻¹ day⁻¹; contribution of TP in feces is 0.5 g·person⁻¹ day⁻¹; ᵇTN after 20% loss of nitrogen in ammonia.

The TN loads from the Upper Worka were higher compared with the other sub-catchments. The 'human loads' from the Derekwonz sub-catchment were the least of all catchments.

4.3.5 TN and TP source apportionments

Dominant land use in both catchments are crops and livestock grazing (Table 4.2.; Figure 4.2.). Water flows at the two catchment outlets varied markedly between the two sampling periods (values in C2 > C1; Table 4.5.). For both Leyole and Worka river catchments, TN and TP diffuse source contributions were generally higher compared with point sources; however, estimates for point sources apportionment for TN in C1 were higher than for diffuse sources, accounting for 80% ([44/ 55]*100) of the total TN transfers in the Leyole river. In contrast, the diffuse sources during C2 comprised 95% of the totals and higher in both the Leyole and Worka Rivers compared with C1.

The diffuse losses of TN varied considerably between the two periods. For the Leyole River, these were 38 times higher for C2 compared with C1. For the Worka River, diffuse sources during C2 were about five times higher than during C1. Similar results were found for TP, although differences were less striking (Table 4.5.).

Table 4. 5. Source apportionment of daily average loads of nutrients from diffuse and point sources discharged into the Leyole and Worka Rivers

Sources	Leyole river catchment			Worka river catchment		
Catchment area, km^2	17.3			31		
Agricultural land, %	56			52		
Population density, persons km^{-2}	1,193			680		
	C1	C2	Average	C1	C2	Average
Mean daily discharge, L s^{-1}	142	296	212	360	1,320	807
Total transfers, kg day^{-1}						
(NH$_4$ +NH$_3$)–N	5	24	15	21	45	33
NO$_3$–N	14	27	21	59	102	81
TN	55	424	240	451	2,306	1,379
PO$_4$–P	13	36	25	137	173	155
TP	206	195	201	1,038	746	892
Point source discharge^a, kg day^{-1}						
(NH$_4$ +NH$_3$)–N	1	1	1	0.4	1	0.7
NO$_3$–N	2	0.4	1	4	4	4
TN	44.3	21.3	32.8	33	88	61
PO$_4$–P	4	1.3	3	3	3	3
TP	37.8	49.1	43.5	35	28	32
Diffuse sources transfers^b, kg day^{-1}						
(NH$_4$ +NH$_3$)–N	4	23	14	20.6	44	32.3
NO$_3$–N	12	26.6	20	55	98	77
TN	10.7	402.7	207.2	418	2,218	1,318
PO$_4$–P	9	34.7	22	134	170	152
TP	168.2	145.9	157.5	1,003	718	860

Note. Diffuse sources apportionment was calculated by subtracting the TN and TP point sources transfers (Table 5.1.) from total TN and TP transfers, sampling period June 15–September 30, 2013 (C1) and 2014 (C2). TN: total nitrogen; TP: total phosphorus.

[a]Effluents from five factory discharges.

[b]Diffuse nutrient loads from background nutrient transfers (i.e., natural land including forest and bare land), transfer from agriculture activities, transfer from scattered villages' areas, and atmospheric deposition onto open waterbodies.

4.4 Discussion

4.4.1 Nutrient transfers into the streams of the Kombolcha catchments

The study provides the first estimates of nutrients loads in the semiarid and industrializing Kombolcha catchment and is one of the few studies estimating and apportioning loads from diffuse and point sources in the region. While a 15-day sampling frequency likely underestimates concentrations of nutrients subject to possible high variance from mobilization associated with high rainfall events (Fauvel et al., 2016), and possible intermittent chemical emissions from the factories, high intensity monitoring with, for example, auto samplers (Anderson and Rounds, 2010) was not possible because of cost and logistics. Nevertheless, the study provides important information on relative loadings of nutrients and which can guide management to reduce nutrient loads in Kombolcha and similar environments.

The Kombolcha factories, especially the brewery, meat processing, and, for period C2, the textile factory, provided considerable TN and TP discharges into the Leyole and Worka rivers, exceeding, for some or all of the time, emission guidelines (EMoI, 2014) of 40 and 5 mg L^{-1} for TN and TP,[1] respectively (Table 4.1.). The brewery effluent showed higher discharge of TN and TP concentrations compared with available data from other African countries (Table 4.1. and 4.6.). Under low flows, proportional concentration of effluent discharge to the rivers become more important (Halling-Sorensen and Jorgensen, 2008).

Table 4.6. Literature values of TN and TP concentrations in some African brewery effluents

Nutrient (mg L^{-1})	Ethiopia (This study)	South Africa (Abimbola et al., 2015)	Nigeria (Inyang et al., 2012)	Zimbabwe (Parawiraa et al., 2005)
TN	10 - 89	0 - 5.36	0.39	0.0196–0.0336
TP	4.1 - 91	-	0.462	16 – 24

Note. TN: total nitrogen; TP: total phosphorus.

The TN loads from diffuse sources in the Leyole catchment increased by a factor of 21 (409/19)

from 2013 to 2014, with a similar trend evident in the Worka catchment. This large increment likely arises from increased hydrological flows in the catchments in the second year of the study (Table 4.5.). The NO_3–N concentrations at the outlet of Tebissa sub-catchment was highest (3.91 mg $N \cdot L^{-1}$), with proportionally largest grazing land compared with the other catchments. TN concentrations reduced with decreasing proportion of grazing land in the sub-catchments, being lowest in the Derekwonz sub-catchment (Figure 4.3.a). No fertilizer is used on the open grazing lands, but farmers leaving manure on the fields may contribute to large NO_3–N concentrations.

Compared with regions in Europe, ranging from an average of 19 kg $N \cdot ha^{-1}$ in Portugal to 125 kg $N \cdot ha^{-1}$ in The Netherlands (Velthof et al., 2014), fertilizer N inputs to the croplands of the Kombolcha area appear quite high, averaging 46 kg $N \cdot ha^{-1}$ of urea ammonium nitrate (i.e., $(CO(NH_2)_2)$–(NH_4NO_3)) and 22 kg $N \cdot ha^{-1}$ DAP $[(NH_4)_2PO_4]$ per year (Kombolcha Agricultural Office, 2015). The export of TN from human defecation from the Upper Worka catchment was comparable with TN loads from the meat processing factory (Tables 4.1). However, although the estimated human diffuse loads may be overestimated because not all people will practice open defecation, it is common throughout sub-Saharan Africa, and waste collected in tanks is often untreated (Rose et al., 2015; Corcoran et al., 2010). Ethiopia has one of the highest number of people openly defecating in the world (WHO/UNICEF, 2014). There is clearly a need to further assess impacts not only for nutrient loads but also for public health. The much higher loadings of particulate compared with dissolved P in all sub-catchments (Figure 4.4.) likely reflects the prevalence of land degradation, especially associated with high slope landforms used for cultivation and livestock grazing. Visible erosion occurs throughout the catchments and is especially prevalent among the steeper hills, with slopes as high as 40%. This can account for the highest TP transfers occurring in the Upper Worka sub-catchment (3.8 mg $P \cdot L-1$) that, while containing the largest percentage (27.6%) of forest lands of all sub-catchments studied (Table 4.2.), is also the most hilly. Ambo sub-catchment is relatively flat.

The significant difference in the concentrations of PO_4–P while comparing the upstream sites with their corresponding catchment outlets (Table 4.3.) shows the presence of other forms of phosphorus in the catchment outlets, most likely as particulate phosphorus considering the soil erosion problem in the area. Besides, the concentrations of TN were significantly higher in the outlets of Worka catchment compared with the river's upstream site, showing the clear influence of land uses and factory emissions on the downstream water quality. In the downstream sections of the Leyole and Worka rivers, the water quality, especially for irrigation and livestock supply, would fail commonly used environmental quality standards for TN and TP (Figure 4.5.).

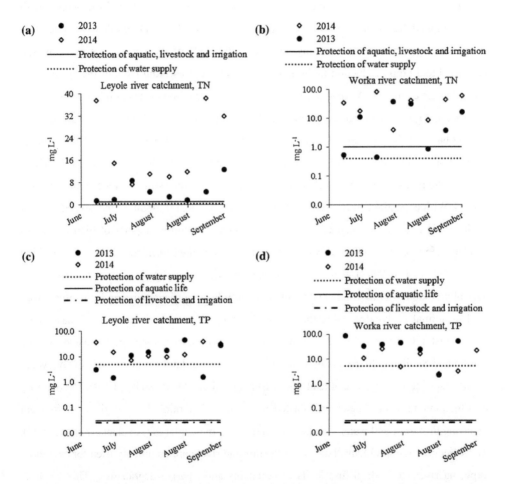

Figure 4.5. Total nitrogen (TN) and total phosphorus (TP) concentrations (mg L−1) at the outlets of the Leyole (LO) and Work rivers (WO) river catchments, compared with the guidelines for protection of aquatic life (Water quality guidelines for the protection of aquatic life, TN: 1 mg N·L−1, (Alberta Environmental Protection, 1993); TP: 0.03 mg P·L−1, (Macdonald et al., 2000a), human 6 water supply (water quality guidelines for the protection of human water supply, TN: 0.4 mg N·L−1, Japanese Water Quality Bureau [JWQB], 1998; TP: 5 mg P·L−1, (Whitehead, 1988), livestock water supply (water quality guidelines for the protection of livestock water supply, TN: 1 mg N·L−1, (JWQB, 1998); TP:0.025 mg P·L−1) (RIDEM, 1997), and irrigation (water quality guidelines for the protection of irrigation, TN: 1 mg N·L−1, (JWQB, 1998); TP: 0.025 mg P·L−1, (JWQB, 1998)

Industrial expansion planned for Kombolcha City will inevitably lead to further degradation of surface waters unless appropriate and locally supported pollution control measures are put in place. As Ethiopia and, in general, African countries strive to improve production and economic opportunity, it is tempting to develop now and deal with environmental and economic consequences later. This strategy is increasingly seen as a false economy (Rudi et al., 2012). The release of nutrients into surface water can affect water quality for irrigation, with associated land degradation (Nyenje et al., 2010; Arimoro et al., 2007). In Ethiopia, agriculture is the leading sector in the economy accounting for 43% of the country's gross domestic product. Field crop production is the major sub-sector representing 64% of the agricultural gross domestic product (Awulachew et al., 2010). Increased food production is a primary goal of Ethiopian government policy, promoting the use of fertilizers to enhance crop production. On the basis of the Ethiopian Central Statistical Agency of 1994/1995–2005/2006, average application rates of urea and DAP are 31 and 16 kg $N \cdot ha{-}1$, respectively (Endale, 2011). These values belie regional and local differences and are less than the estimates for Kombolcha (see above). With expanding sizes of farms, and a growing national floriculture industry, the use of fertilizers is expected to increase substantially, adding to current reports of high emissions (Endale, 2011; Getu, 2009). Although Ethiopia has regulated fertilizer storage and packing since 2002 (EEPA, 2002), there are no policies restricting application rates of fertilizers to agricultural lands, or environmental regulation controls on nutrient emission from either cropland or animal husbandry. More than 85% of Ethiopian farmers operate on <2 ha of land (Economist Intelligence Unit, 2008), with concomitant pressure for more intensive farming (Smith and Siciliano, 2015). As in many African countries, land tenure is exclusively owned by the State, reducing a sense of land stewardship (Taddese, 2001; Gavian, 1999). The majority of Ethiopian farmers depend on rain-fed agriculture, but rainfall is erratic with frequent recurring droughts, and water storage is inadequate in many places (Awulachew et al., 2007). To overcome these problems, lands near streams and rivers are commonly used for irrigation, exposing rivers to high diffuse loads of pollutants and riparian degradation. There is little discussion on protection of rivers using riparian buffer zones.

Ethiopian geography varies greatly, ranging from high peaks of 4,550 m above sea level to a low depression of 110 m below sea level. Government efforts to reduce soil erosion rely on constructing terraces on high gradient lands. However, poor design and common lack of maintenance result in continuing high rates of erosion, accentuated by overgrazing by livestock and deforestation (Tefera and Sterk, 2010; UNEP, 2004). Particulate phosphorus and nitrogen transported with soil particles are important sources stimulating eutrophication of surface waters. Ethiopia has among the largest livestock population density in Africa (Negassa and Jabbar, 2008), with permanent grazing land constituting about 20 million hectares, or 20% of the total land use cover of the country (African Development Bank, 2017). Livestock feeding is mostly uncontrolled with free grazing across communally owned pasturelands (Kebede, 2002). Where densities of livestock are high, soil compaction increases surface run-off and mobilization of livestock manure, leading to nutrient loads to surface waters (McDowell, 2008). Satellite imagery shows that land degradation hotspots over the last three decades comprise about 23% of the land area in the country (Gebreselassie et al., 2016).

This preliminary study has identified some key issues important for future assessment and management of nutrient loads in cities such as Kombolcha, providing (a) quantitative data on the effluent P and N concentrations discharged from the local factories; (b) upstream/ downstream trends of nutrient concentrations and relationship with land uses; and (c) estimation of the apportionment of nutrient loads of P and N between point and diffuse sources.

Further monitoring will need to balance collecting the necessary information with limited resources. Realistically, routine monitoring is limited to the rainy season (June–September) in Kombolcha. Sampling frequency in this study was approximately fortnightly, with no possibility for deployment of continuous automated samplers. In The Netherlands, driven by an expanding list of monitoring variables (>250 in 2018 [www. rijkswaterstaat.nl/water]), together with appreciable budget restrictions, a number of monitoring optimization programmes have been carried out over the last decades. Thus, Ottens et al. (1997), for North Sea monitoring, concluded that monitoring frequencies less than once per month offered acceptable levels of trend detection (about 15%), with only a relatively small improvement, to about 10%, at bi-monthly frequencies. On the basis of this and other studies, The Netherlands monitoring programmes for 'Governmental waters' have reduced routine monitoring frequencies from, generally, once per week in the 1980s to once per month now (for details, see www.rijkswaterstaat.nl/water).

Sediment loss and diffuse nutrient pollution from poor land use management remain a global concern. Although techniques for wise use of soil and fertilizer management are well developed, these are often supported with extension services designed to optimize crop growth and minimize nutrient and soil losses. Ethiopia could very much benefit from such a service, requiring a supporting policy framework and political will to promote sustainable agriculture and land use. Ethiopia has signed up to the Sustainable Development Goals (UNDP, 2015) and the Paris Climate Agreement (United Nations, 2016). This commitment can only work if translated to a local response. The city of Kombolcha can be a good test of that.

4.5 Conclusions

The Ethiopian city of Kombolcha exemplifies the challenges of measuring and managing nutrient emissions from land use and industrial sources that occurs across sub-Saharan Africa. Estimates of nutrient loads over two sampling programmes identified the importance of diffuse loads from land and point sources from some of the industrial units in the city. Poor land management plays a major role in this. High percentage of crop lands were associated with increasing nitrogen concentrations in the rivers, but soil erosion from upland forested areas probably make a substantial contribution to TP loads. Industrial nutrient pollution of concern was found from a local meat factory and brewery. Although these industries contributed relatively low amounts of the total nutrient loads during the wetter of the study years, during the drier year, with lower flows, the relative contribution was substantial. The management of such point sources to reduce total emissions is straightforward, and pollution from the factories can be largely eliminated with effective licensing. Nutrient concentrations arising from the upstream sub-catchments were variable and increased downstream. Future development plans for industry and agriculture present a major risk for surface water quality. Human resources, expertise, and infrastructure remain a major gap in monitoring, so attention to greater, and verifiable, use of land and water quality models is of major importance to guide monitoring and management.

Chapter 5

Estimating total nitrogen and phosphorus losses in a data-poor Ethiopian catchment

Publication based on this chapter:

Zinabu E, van der Kwast J, Kelderman P, Irvine K. 2017. Estimating total nitrogen and phosphorus losses in a data-poor Ethiopian catchment. Journal of Environmental Quality, 46:1519-1525.

Abstract

Selecting a suitable model for a water quality study depends on the objectives, the characteristics of the study area and the availability, appropriateness and quality of data. In areas where in-stream chemical and hydrological data are limited, but where estimates of nutrient loads are needed to guide management, it is necessary to apply more generalized models that make few assumptions about underlying processes. This paper presents the selection and application of a model to estimate Total Nitrogen (TN) and Total Phosphorus (TP) loads in two semi-arid and adjacent catchments exposed to pollution risk in north-central Ethiopia. Using specific criteria to assess model suitability resulted in the use of PLOAD. The model relies on estimates of nutrient loads from point sources such as industries and export coefficients of land use, calibrated using measured TN and TP loads from the catchments. The performance of the calibrated PLOAD model was increased, reducing the sum of errors by 89 % and 5 % for the TN and TP loads, respectively. The results were validated using independent field data. Next, two scenarios were evaluated: (1) use of riparian buffer strips, and (2) enhanced treatment of industrial effluents. The model estimated that combined use of the two scenarios could reduce TN and TP loads by nearly 50%. Our modelling is particularly useful for initial characterization of nutrient pollution in catchments. With careful calibration and validation, PLOAD model can serve an important role in planning industrial and agricultural development in data-poor areas.

5.1 Introduction

Catchment-based water quality models are essential for management of water quality in industrialized and urbanized catchments (Álvarez-Romero et al., 2014; Wang et al., 2013). In the developing world, however, reliable application of water quality models is often lacking (Wang et al., 2013; Reggiani and Schellekens, 2003; Singh, 1995) for three important reasons. Firstly, limited human capacity to use modelling software hampers the use of water quality modelling in water quality management (Rode et al., 2010; Loucks et al., 2005). Secondly, the availability and quality of data is often inadequate and lastly, access to proprietary software and decision support systems is limited by finances.

Filling these gaps requires both development of local human and institutional capacity, as well as design of programmes for data collection and monitoring. While there can be temptation to invest in quite complex modelling, this does not necessarily result in more accurate understanding of the underlying processes on which such models are based. Such models can also be costly and subject to large errors in predictions from deficiencies in the data (Ongley and Booty, 1999). Therefore, starting with a basic model and gradually employing more detailed and comprehensive models is a sensible approach. Low cost and less complex models that do not require extensive data sets are useful general approaches for the prediction of, particularly, diffuse, pollution from land use (Ding et al., 2010; Johnes, 1996) and industrial development (Gordon, 2005); and for assessing likely results from different management scenarios such as grass buffer strips (Dorioz et al., 2006). A common approach in such situations has been the use of generalised export coefficients that predict an annual load of nutrients from land to water. While such a "black-box" approach may lack insight into underlying hydrological or chemical processes that vary with precipitation and terrain (Noto et al., 2008), they have been useful in estimating overall catchment loads and, at least, relative effects under different land uses (Shrestha et al., 2008; Ierodiaconou et al., 2005; Soranno et al., 1996). Using an export coefficients approach can be especially advantageous for data-poor areas and for initial estimates relating land use to water quality (Bowes et al., 2008).

In northern central Ethiopia, Kombolcha City is developing as an industrial hub. The city lies within a varied landscape of agricultural activities in the rural uplands, and a largely urbanized and industrial lowland. Downstream sections of the rivers supply water for irrigation. Although Ethiopia adopts the WHO guidelines for drinking and irrigation water quality (Ademe and Alemayehu, 2014), water quality control or river monitoring are barely implemented. Existing industries in Kombolcha have limited treatment of waste water and, together with already moderately intensive land use, provide nutrient pressures to a small river network. Planned development of the city, and further intensification of surrounding land risks increase of nutrient loads to the rivers flowing through the city. As in many similar situations in both Ethiopia and Sub-Sahara Africa, local authorities lack the capacity to either monitor or predict ambient river nutrient concentrations or loads and, therefore, suffer from a severe knowledge deficit to guide sustainable development. Use of simple water quality models is one way to address this deficit, but requires better decision making in identifying which models are likely to be of cost-effective benefit.

In this study, we examined the feasibility of a number of water quality models applicable to the semi-arid landscape and industrial activities of Kombolcha City. Specific objectives were to: (i) screen a number of models for their applicability to the semi-arid and data poor regions, (ii) estimate the annual TN and TP loads from the Kombolcha's catchments using the applicable model; and (iii) Simulate the change in the TN and TP loads due to best management practices (BMPs) and enhancing the efficiency of point sources treatment facilities.

5.2 Material and methods

5.2.1 Study site description

Kombolcha, in the North central part of Ethiopia, is bordered by moderately intensive agriculture, grasslands and natural and plantation forest lands in the north-west and south-west, and industrial areas in the center and north-east (Zinabu et al., 2018). The semi-arid climate has a rainy season from mid-June to the end of September. The landform comprise high plateaus, the Borkena graben and southward sloping ground to the Borkena River. Soils throughout the catchment are dominated by Vertisols (Zinabu, 2011). High soil erosion, often resulting in deep gullies, occurs through the catchment. The naturally ephemeral Leyole and Work rivers that flow through the city pass through and receive effluent from the industrial areas before entering the larger Borkena River. Additionally, the urban area receives diffuse loads from upstream agricultural and forested areas. Teff [*Eragrostis tef (Zuccagni) Trotter*] and maize (*Zeamays L.*) are typical rainfed crops growing in the rural areas, whereas tomatoes (*Solanum Lycopersicum L.*) and lettuce (*Lactuca sativa L.*) grow under irrigation in the upstream areas. Most farmlands are dissected by gullies, and high soil erosion is evident.

5.2.2 Model selection

An initial screening of feasible models to estimate loss of nutrients from land to water was predicated on assessment of a) relatively simple process description and adaptability for data-poor areas where at most simple quantification of land cover characteristics may exist; b) the availability of data for estimation of model inputs and parameters; and c) free and/or open source software (FOSS) given budgetary limitations for future operational use. A range of physical and empirical models were screened based on this set of criteria (Table 5.1.). This systematic approach identified the PLOAD model (incorporated in the BASINS 4.1 System) as the most suitable.

Estimating total nitrogen and phosphorus losses in a data-poor Ethiopian catchment

Table 5.1. Modelling criteria evaluation matrix for proposed modelling tools, "Yes" depicts fulfilment of criteria and "No" shows desertion of criteria by a model tool

Model tool	Based on export coefficients	Spatial discretization		Annual Temporal scales	Complexity		Not high data demanding	Public accessibility		Multi-functionality (tables and graphics outputs)	Domain of model application (River model vs. catchment process model)
		Distributed†	Semi-distributed‡		Not process based	Adaptable to unique conditions		Free ware	Open source		
AGNPS (Karki et al., 2017)	No	No	Yes	Yes	No	Yes	Yes	Yes	Yes	Yes	Catchment
GWLF (Haith et al., 1992)	Yes	No	Yes	Yes	No	Yes	Yes	Yes	Yes	Yes	Catchment
HSPF (Bicknell et al., 1993) (in BASINS 4.1 System)	No	Yes	No	Yes	No	Yes	Yes	Yes	Yes	Yes	Catchment/River
MIKE 11 (DHI, 1998)	Yes	Yes	No	Yes	No	No	Yes	No	No	Yes	River
MONERIS (Behrendt et al. 2007)	Yes	No	Yes	Yes	Yes	Yes	Yes	Yes	Yes	Yes	River
PLOAD (USEPA, 2015) (in BASINS 4.1 System)	Yes	No	Yes	Yes	Yes	Yes	Yes	Yes	Yes	Yes	Catchment
PolFlow (De Wit, 1999)	No	Yes	Yes	Yes	Yes	Yes	No	Yes	No	Yes	Catchment
QUAL2E (Brown and Barnwell, 1987)	No	No	No	Yes	No	Yes	Yes	Yes	No	Yes	River
SIMCAT (Warn, 2010)	No	No	No	Yes	Yes	No	Yes	No	Yes	Yes	River
SPARROW (Schwarz et al., 2006)	Yes	No	Yes	Yes	Yes	Yes	Yes	No	No	Yes	Catchment/River
SWAT (Arnold et al., 1998)	No	No	Yes	Yes	No	Yes	No	Yes	Yes	Yes	Catchment/River
TOMCAT (Bowden and Brown, 1984)	No	No	No	Yes	Yes	No	Yes	No	Yes	Yes	River
WASP (Ambrose et al., 1988)	No	No	No	Yes	No	Yes	No	Yes	Yes	Yes	River

† models that perform all calculations on a grid-base and then route the water flow through the model domain until a subcatchment and/or catchment outlet

‡ models that divide the subcatchments into sub-areas based on land-use, soil and/or other information to form "response units" or "contributing areas"

BASINS 4.1 system and PLOAD model

BASINS (*Better Assessment Science Integrating Point and Non-point Sources*) is a customized open source GIS application (*MapWindow*) designed for watershed and water quality based studies (USEPA, 2015, 2008; Edwards and Miller, 2001). The current version "BASIN" 4.1" system incorporates the catchment modelling tools HSPF, SWAT, PLOAD and SWMM (USEPA, 2015). The GIS feature in BASINS provide a visual interpretation of data and displays landscape information, thus allowing to map and integrate land use and point source discharges at a scale chosen by the use.

PLOAD estimates non-point nutrient loads, and can be applied for urban, suburban and rural areas (USEPA, 2015; Young, 2010; Lin and Kleiss, 2004).The model can be run for multiple scenarios. Results are reported in such a way that different scenarios associated with reduction of pollutant loads can be easily compared (Gurung et al., 2013; Edwards and Miller, 2001). Supplemental S2-3 shows the methods and algorithms that are used to calculate pollutant loads in the PLOAD.

PLOAD Model concept

The PLOAD model calculates pollutant loads from mixed land use in catchments using either of two algorithms: "Event mean concentrations" or "Export Coefficient " (USEPA, 2015; Edwards and Miller, 2001). Both methods can be applied under different circumstances, the available data and site characteristics are amongst important factors. Based on the available data, we found the "Export Coefficient" more applicable in our study area (Table 5.2.). The export coefficient model achieves acceptable accuracy when adopted on areas predominantly with agricultural activities whereas, the simple method is used mainly for urban scales (Edwards and Miller, 2001).

The export coefficients are distinct values for the characteristics of a particular land-use runoff and it is a measure of estimated mass (for. e.g.TN or TP kg) loss per unit area per year. The export coefficient method basically requires two types of input data: land-use maps and all unit loads of land use (i.e. land-use export coefficient values) (Edwards and Miller, 2001). The land-use maps provide the area of each land-use type and export coefficient values will be assigned to each land use. The specific pollutant loads are calculated using Equation 1:

$$L_P = \sum_U (L_U \times A_U) \qquad\qquad Equation\ 1$$

Where: L_P is the pollutant load (kg $year^{-1}$);

L_U is pollutant loading rate for land $-$ use type U $\left(\dfrac{kg/km^2}{year}\right)$; and

A_U is area of land $-$ use type u (km^2).

Table 5.2. Comparison of the availability of input data for Kombolcha's catchments in order to select either the "*Simple Method*" or "*Export Coefficient*" methods in PLOAD modelling process

Event mean concentrations		Export coefficient method	
Model input	Data availability	Model input	Data availability
Catchment boundaries	Yes	Catchment boundaries	Yes
Land use coverage	Yes	Land use coverage	Yes
Annual precipitation	Yes	Export coefficient	From literature, (Kato et al., 2009; Shaver et al., 2007; Lin, 2004; Johnes, 1996)
Event mean concentration	From literature, (Kato et al., 2009; Packett et al., 2009; Shaver et al., 2007; Lin, 2004)	BMP	From literature: e.g. (Horst et al., 2008; Shaver et al., 2007)
Land use imperviousness	Yes	Point source pollutant	Yes
BMP	From literature (Horst et al., 2008; Shaver et al., 2007)		
Point source pollutant	Yes		

The loading rates were derived from the export coefficient for each land use, while the land-use areas were interpreted from the land use and catchment GIS data. Here, the pollutant loads derived from this methods were refined to include loads from point sources and also investigate the remedial effects of BMPs. Two equations (Equations 2 and 3) were used to in the PLOAD model to recalculate the pollutant loads for a catchment serviced by BMPs such as riparian buffer strips (Edwards and Miller, 2001)::

i.   The pollutant loads remaining after removal by each BMP (i.e. grass buffer) were calculated using equation 3:

$$L_{BMP} = (L_P \times \%AS_{BMP}) \times \left(1 - \dfrac{\%EFF_{BMP}}{100}\right) \qquad Equation\ 2$$

Where: L_{BMP} is grass buffer loads (kg)

L_P is raw catchment loads (kg)

AS_{BMP} is percent area serviced by grass buffer (decimal percent)

EFF_{BMP} is percent load reduction of grass buffer (%)

Here, the raw pollutant loads were derived from the calibrated PLOAD model, whereas the percent load reduction was taken from literature values for the BMPs of grass buffers (Horst et al., 2008; Shaver et al., 2007). For acceptable nutrient management, Horst et al. (2008 recommends that the load reduction efficiency need to be 40% and 45% for TN and TP, respectively. This is based on North America catchments and there transferability to Kombolcha's condition might be a limitation. However, the data are unavailable and hard to measure the pertinent local conditions, and we used results from a comparable catchment. Each loads reduction by the grass buffer practices were then calculated from the full pollutant load coming off the catchment.

ii.   The total pollutant loads accounting for BMPs were determined by catchment using the equation presented below. Each catchment load was a cumulative total of areas that were and were not covered by BMPs.

$$L = (\textstyle\sum_{BMP}(L_P)) + L_P \times (A_B - (\textstyle\sum_{AS} \times (AS_{BMP}))) \qquad \text{Equation 3}$$

Where: L is loads after grass buffer (kg)

L_P and AS_{BMP} are as described in the previous, equation (2)

A_B is area of catchment (km^2)

Data needs and PLOAD model processing

We used the "Export Coefficient" option (USEPA, 2015; Edwards and Miller, 2001) in PLOAD to estimate the potential nutrient loads from mixed land use. To estimate diffuse pollution, each land-use category was assigned TN and TP export coefficient values (kg ha^{-1} year^{-1}). As these coefficients are not available for Ethiopia, we used values following a review of the literature for applicable land uses (Table 5.3.) (Ding et al., 2010; Wood and Beckwith, 2008; Yetunde, 2006; Lin, 2004; EPA, 2001). The land use and topography of Kombolcha, comprises areas of forest, grassland and crops on hill sides, mountainous areas and relatively flat lands (Zinabu et al., 2018). Use of applicable coefficients were further informed from consultations with the local Bureau of the Water and Agricultural Office on soil, topography, imperviousness, and vegetation cover. Details of the methods used to measure water flows in the river are reported by Zinabu et al. (2018). Nutrient concentrations in the Leyole and Worka rivers were measured in two monitoring campaigns when there was sufficient water flowing from 15 June to 30 September in both 2013 (C1) and 2014 (C2). Nutrient loads from the factories were estimated using nutrient concentrations and effluent discharge rates from outlet pipes, while nutrient loads from the catchments were derived using nutrient concentration and stream discharges at the catchments' outlets. Total loads to a catchment outlet were computed using the Flux 32 software (Walker, 1999).

Table 5.3. Published Export coefficients of TN and TP, in kg ha^{-1} yr^{-1}, for various land uses and the selected export coefficients values for the PLOAD model TN and TP loads estimations

Land use	Export Coefficients			
	Ranges in Literature[§]		Selected for modelling	
	TN	TP	TN	TP
		kg ha^{-1} yr^{-1}		
Water body	0.69 – 3.8	0.09 – 0.21	0.75	0.2
Bare lands	0.5 – 6.0	0.05 – 1.13	5.6	0.25
Forest land	1 – 6.3	0.007 – 1.11	3.4	0.8
Grass land	3.2 - 14	0.05 – 18.61	13.5	0.3
Industrial area	1.9 - 14	0.4 – 7.6	13.5	3
Residential area	5 – 7.3	0.77 – 2.21	6.1	2.1
Crop land	2.1 - 79.6	0.06 – 18.61	79.5	2

5.2.3 Calibration and validation methods

After setting up the PLOAD model, export coefficients were calibrated. The performance of the PLOAD model was assessed by measuring the percentage of error of estimation Equation 4, comparing loads estimated by the PLOAD model with those derived from nutrient measurements of 2013 (Table 5.4.). The sum of the percentage errors from all the six sub-catchments were used to calibrate the export coefficient values of the land use used in the PLOAD model. To optimize export coefficients of the land use in the catchment, the goal is to minimize the model's sum of errors. The Microsoft Excel (2013) Solver was used in finding the smallest sum of the percentage errors. In this study, a value of zero was set for the sum of the absolute percentage errors in the objective cell and the export coefficients were set to be optimized in the Solver function to retain the objective value (i.e. zero) or closest to it. Lower and higher limits were selected for the export coefficients value of each land use from literature and these limits were used to constrain the changes of the export coefficients in the Solver function to avoid unrealistic values of export coefficients (see section "Export coefficients of land use"). In the Solver, the GRG nonlinear method was used for the optimization of the export

---

[§] Source: Loehr et al. (1989), Lin (2004), (Yetunde 2006), (USGS 2008), (EPA 2001)

coefficients because at least one of the input parameters (i.e. export coefficients) is assumed to be a nonlinear function of the land-use variables in the PLOAD model (Fylstra et al., 1998).

Table 5.4. Mean daily flows and TN and TP loads from factory effluents and catchments, estimated for the monitoring campaigns C1 (2013) and C2 (2014); n=8 for each year; source: Zinabu et al. (2017)

	Sources	Mean daily flows		TN		TP	
		C1	C2	C1	C2	C1	C2
		L s^{-1}		kg year^{-1}			
Factory	Brewery	8.2	21	12,045	32,120	12,770	10,220
	Meat Processing	11	8.8	13,140	5,480	13,500	10,220
	Textile	15	16	1,940	1,420	182	5,800
	Tannery	6.8	8.4	913	694	117	1,500
	Steel Processing	1.7	2.2	183	183	15	401
Catchment	Derekwonz	6	14	523	962	870	571
	Ambo	24	37	2,960	7,200	9,160	2,880
	Tebissa	27	57	1,360	6,470	7,160	16,570
	Leyole River	142	296	3,900	146,990	61,320	53,290
	Upper Worka	360	1,320	41,430	103,290	77,230	221,890
	Worka River	360	1,320	152,570	809,570	366,100	262,070

The optimized export coefficients vary for each catchment, and therefore, a central value that results in lower errors is the feasible option to be used in the model. To find the central value, the median, average and weighted average of these optimized export coefficient values of the catchments were tested to choose values (i.e. export coefficient of each land-use category) that calibrate the model with lowest sum of error percentages. Other dependencies factors affecting export of TN and TP are hard to apply, as related data such as soils, hydrology and topography are scant and problematic to measure in the areas (Zinabu et al., 2018).

$$Error\ of\ estimation\ =\ \left(\left(\frac{Monitored\ loads-PLOAD\ estimation}{Monitored\ loads}\right)\times 100\right) \qquad \text{Equation 4}$$

The calibrated model was validated using the measurements from 2014 (Table 5.2.).

5.2.4 Definition of scenario

Two scenarios were defined:

- Enhancing the efficiency of the factory effluent treatments: we predicted the change in the TN and TP loads for a 50% increment in the efficiency of the effluent treatment facility. The monitoring dataset from the factory effluent were used to derive the concentrations of TN and TP loads accounting for the enhanced efficiency of the treatment by calculating the 50 % reduced loads.

- Practicing riparian grass buffer strips: to predict the influence of the riparian buffer strips, we first estimated the nutrients loads using the PLOAD export coefficient method and then the loads with inclusion of grass buffers in each catchment. For this study, we were guided by Horst et al. (2008), who recommended 40% and 45% for TN and TP target for nutrient management. Equations used in the PLOAD model to recalculate the pollutant loads serviced by BMPs are shown in Electronic Supplementary Material (ESM).

The results of these scenarios were compared with the current status of nutrient loads in the Leyole and Worka catchments.

5.3 Results

5.3.1 Calibration of the PLOAD model

The calibration method resulted in a reduction of the sum of errors to 648 % compared with 5,390 % in the pre-optimized model (Table 5.5.a, c). Using the average and weighted average values of these export coefficients, the sum of error of estimations were reduced to 1,450 % and 2,390 %, respectively. However, the median values resulted in a lower sum of error of estimation (648 %). For the individual catchments, the biggest error of estimation was found for the Tebissa sub-catchment, while a smaller error of estimation was found for the Ambo sub-catchment (Table 5.5.c). The fully calibrated PLOAD model often resulted in a larger error of estimation for the larger catchments than sub-catchments. In contrast, for the TP loads, the sum of error of estimations pre-optimizing the export coefficients was 373% (Table 5.6.a). After optimizing the export coefficients using the *Solver* function, the sum of errors in the model was reduced to 296 % (Table 5.6.b).

The sum of the errors using the median and weighted average values of the optimized export coefficients, for each land use in each of the catchments, was 366% and 363%, respectively (Table 5.6.c), which is barely reduced compared with the pre-optimized sum of errors (373%). However, a more reduced sum of errors (356%) was found while using the average values of optimized export coefficients. The errors of estimation show larger errors in the catchments than the sub-catchments, with the smallest error for the Tebissa sub-catchment and largest error for the Worka catchment.

Table 5.5. Measured TN loads of catchments and estimated loads by the PLOAD model and error of estimation (the direction of the error depicted as "(+)" for overestimation, while "(-)" is underestimation) and sum of the absolute error of estimation for TN export coefficients of land use in the study areas for the case of: a) non-optimized export coefficients; b) optimized export coefficients using MS Excel Solver add-in; c) calibrated PLOAD model using the median values optimized export coefficients; and d) validation results of the calibrated PLOAD model; the TN loads were based on the dataset of the monitoring campaign in 2014 (C2)

Catchment	Export coefficients (kg ha⁻¹ year⁻¹)							PLOAD estimation	Measured loads	Error of estimation (%)
	water body	Bare land	Forest land	Grass land	Industrial area	Residential area	Crop land			
								Kg year⁻¹		
(a) Nonoptimized export coefficients										
Derekwonz	0.75	5.6	3.4	13.5	13.5	6.1	79.5	6,000	523	(-) 1,047
Ambo	0.75	5.6	3.4	13.5	13.5	6.1	79.5	14,690	2,960	(-) 397
Tebissa	0.75	5.6	3.4	13.5	13.5	6.1	79.5	32,260	1,360	(-) 2,267
Leyole	0.75	5.6	3.4	13.5	13.5	6.1	79.5	62,720	3,900	(-) 1,506
Upper Worka	0.75	5.6	3.4	13.5	13.5	6.1	79.5	99,430	41,400	(-) 140
Worka	0.75	5.6	3.4	13.5	13.5	6.1	79.5	102,000	152,600	(+) 33
Sum of errors										5,390
(b) Optimized export coefficients using the Microsoft Excel Solver										
Derekwonz	2.43	4.06	4.17	10.05	8.79	6.42	3.32	523	523	0
Ambo	1.50	1.24	3.41	8.71	9.22	6.06	14.32	2,960	2,960	0
Tebissa	2.02	0.51	1.02	3.21	5.09	5.00	2.10	1,540	1,360	(-) 13
Leyole	2.52	1.04	1.35	4.22	3.44	6.62	2.16	3,900	3,900	0
Upper Worka	1.54	0.54	1.16	5.57	10.76	5.67	34.11	41,430	41,400	0
Worka	1.61	6.00	6.13	13.85	6.77	6.69	79.50	104,200	152,600	(+) 32
Median	1.81	1.14	2.38	7.14	7.78	6.24	8.82	-	-	-
Average	1.84	2.23	2.87	7.60	7.35	6.08	22.6	-	-	-
Weighted average	2.07	2.47	3.31	7.79	4.79	5.82	37.48	-	-	-
Sum of errors										45
(c) Calibrated PLOAD model using the median value optimized export coefficients										
Derekwonz	1.81	1.14	2.38	7.14	7.78	6.24	8.82	779	523	(-) 49
Ambo	1.81	1.14	2.38	7.14	7.78	6.24	8.82	1,920	2,960	(+) 35
Tebissa	1.81	1.14	2.38	7.14	7.78	6.24	8.82	4,880	1,360	(-) 258
Leyole	1.81	1.14	2.38	7.14	7.78	6.24	8.82	9,960	3,900	(-) 155
Upper Worka	1.81	1.14	2.38	7.14	7.78	6.24	8.82	15,800	41,400	(+) 62
Worka	1.81	1.14	2.38	7.14	7.78	6.24	8.82	16,300	152,600	(+) 89
Sum of errors										648
(d) Validation results of the calibrated PLOAD model										
Derekwonz								779	962	(+) 19
Ambo								1,920	7,198	(+) 73
Tebissa								4,880	6,471	(+) 25
Leyole								9,960	103,300	(+) 85
Upper Worka								15,800	147,000	(+) 93
Worka								16,310	809,570	(+) 98
Sum of errors								-	-	393

Table 5.6. Measured TP loads of catchments and estimated loads by the PLOAD model and error of estimation (the direction of the error depicted as "(+)" for overestimation, while "(-)" is underestimation) and sum of absolute error of estimation for TP export coefficients of land use in the study areas for the case of: a) non-optimized export coefficients; b) optimized export coefficients using MS Excel Solver add-in; c) calibrated PLOAD model using the average values optimized export coefficients; and d) validation results of the calibrated PLOAD model; the TP loads were based on the dataset of the monitoring campaign in 2014 (C2)

Catchment	Export coefficients (kg ha⁻¹ year⁻¹)							PLOAD estimation	Measured loads	Error (%)
	water body	Bare land	Forest land	Grass land	Industrial area	Residential area	Crop land	kg year⁻¹		
(a) Nonoptimized export coefficients										
Derekwonz	0.21	0.53	0.66	17.14	4.15	1.84	15.90	1,400	870	(-) 56
Ambo	0.21	0.53	0.66	17.14	4.15	1.84	15.90	3,400	9,160	(+) 63
Tebissa	0.21	0.53	0.66	17.14	4.15	1.84	15.90	8,700	7,160	(-) 22
Leyole	0.21	0.53	0.66	17.14	4.15	1.84	15.90	16,670	61,390	(+) 73
Upper Worka	0.21	0.53	0.66	17.14	4.15	1.84	15.90	25,830	77,230	(+) 67
Work	0.21	0.53	0.66	17.14	4.15	1.84	15.90	26,550	366,100	(+) 93
Sum of errors										373
(b)Optimized export coefficients using the Microsoft Excel Solver										
Derekwonz	0.21	0.46	0.52	0.51	5.92	0.99	11.89	870	870	0
Ambo	0.21	1.03	1.11	1.37	1.14	1.05	18.61	3,360	9,160	(+) 63
Tebissa	0.21	0.34	0.23	2.84	1.66	1.16	18.33	7,160	7,160	0
Leyole	0.21	0.09	1.11	18.61	7.11	1.17	18.61	19,122	61,390	(+) 69
Upper Worka	0.21	0.12	1.11	4.45	3.10	1.66	18.61	23,430	77,230	(+) 70
Work	0.21	0.10	0.69	18.61	4.33	1.50	11.59	22,180	366,100	(+) 94
Median	0.21	0.23	0.90	3.64	3.72	1.16	18.47	-	-	-
Average	0.21	0.36	0.80	7.73	3.88	1.25	16.27	-	-	-
Weighted average	0.21	0.17	0.88	11.70	5.98	1.49	16.20	-	-	-
Sum of errors										296
(c)Calibrated PLOAD model using the average value optimized export coefficients										
Derekwonz	0.21	0.36	0.80	7.73	3.88	1.25	16.27	1,270	870	(-) 46
Ambo	0.21	0.36	0.80	7.73	3.88	1.25	16.27	3,120	9,160	(+) 66
Tebissa	0.21	0.36	0.80	7.73	3.88	1.25	16.27	7,290	7,160	(-) 2
Leyole	0.21	0.36	0.80	7.73	3.88	1.25	16.27	14,070	61,320	(+) 77
Upper Worka	0.21	0.36	0.80	7.73	3.88	1.25	16.27	22,140	77,230	(+) 71
Work	0.21	0.36	0.80	7.73	3.88	1.25	16.27	22,770	366,000	(+) 94
Sum of errors										356
(d) Validation results of the calibrated PLOAD model										
Derekwonz								1,270	571	(-)122
Ambo								3,120	2,890	(-) 8
Tebissa								7,290	16,570	(+) 56
Leyole								22,140	221,890	(+) 90
Upper Worka								14,070	53,290	(+) 74
Work								22,800	262,070	(+) 91
Sum of errors								-	-	441

5.3.2 Validation of the PLOAD model

For the TN loads, the PLOAD model performed relatively better for the Derekwonz and Tebissa sub-catchments than the other catchments, with an error of estimation of 19 and 25%, respectively (Table 5.5.d). But, the errors of estimation for TN loads were quite high (98%) for the Worka catchment. For the TP loadings, except for the Ambo sub-catchment, relatively large errors were obtained (Table 5.6.d).

5.3.3 Scenario Outcomes

Total Nitrogen

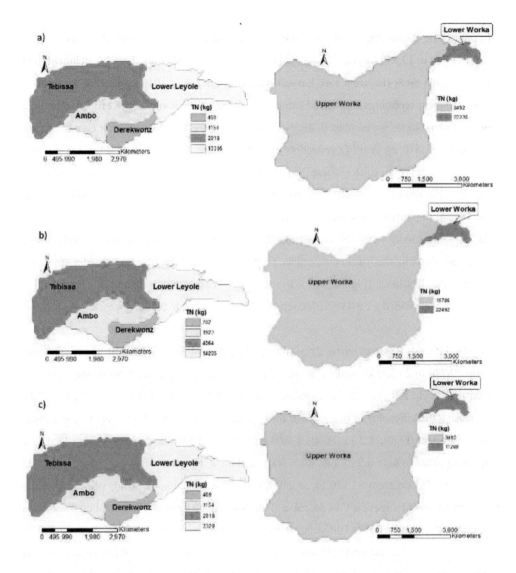

Figure 5.1. Total nitrogen (TN) loads (kg yr−1) for the scenarios of: (a) implementing riparian buffer strips, (b) enhancing of effluent treatment plant, and (c) the joint scenario of a and b in the Tebissa, Ambo, and Derekwonz sub-catchments, and in the Lower Leyole River catchments and the Upper Worka sub-catchment and Lower Worka River catchment.

Though the largest errors were found in the Derekwonz sub-catchment, the error was consistently greater for the larger catchments of the Upper Worka, Worka, and Leyole. The calibrated model results, which represents a baseline situation, was compared with two scenarios: (1) incorporation of riparian buffer strips with nutrient reduction efficiencies of 45% TN and 40% TO; and (2) enhancing the efficiency of treatment of effluents from the factories in the Leyole and Worka catchments by 50% (Figure 5.1. and Figure 5.2.).

In the Leyole catchment, use of riparian buffers provided for a reduction of TN loads by 18%, from 21,760 to 17,840 kg year^{-1} (estimated by summing the loads from the sub-catchments and lower Leyole areas (Figure 5.1.a). Largest scenario decrease of 1940 kg year^{-1} in the Tebissa sub-catchment represents a 40 % reduction. If both buffer strips and a 50% effluent treatment of the four factories are combined, the estimated load in the Leyole catchment could decrease further to 11,870 kg year^{-1} (a reduction of 45%) (Figure 5.1.c). In the Lower Leyole sub-catchment areas (Figure 5.1.), where the four factories are placed, the combined scenarios could considerably reduce the loads from 14,200 to 7,330 kg year^{-1} (48%). In the entire Worka catchment, the buffer strips scenario only reduced the TN loads by 17%, from an estimated 38,280 to 31,830 kg year^{-1} (Figure 5.1.a). With combined buffer strips and enhanced effluent treatment scenario, the loads could reduce to 20,750 kg year^{-1}, and this suggested a 46% total reduction of TN loads (Figure 5.1.c). In the Upper Worka sub-catchment (cf. Figure 5.1.), buffer strips provided for a 45% reduction of TN from 15,790 to 9,490 kg year^{-1}.

For the TP loads Riparian buffer strips indicated a potential for a 21 % reduction of TP loads from 29,770 to 23,520 kg year^{-1} (Figure 5.2.a), in the Leyole catchment, while the combined measures could achieve a 48 % reduction from 29,770 to 15,580 kg TP year^{-1} (Figure 5.2.c). Increasing the efficiency of treatment of the four factories that are found in the Lower Leyole sub-catchment (Figure 5.2.) suggests a 49% reduction in TP loads, decreasing from 18,070 to 9,150 kg year^{-1} (Figure 5.2.b). For the Worka catchment, riparian buffer strips were estimated to reduce TP load from 34,120 to 23,940 kg year^{-1}, a reduction of 30%), while improving effluent treatment achieve a 17% decrease of loads to 28,360 kg year^{-1}. The combined scenarios of buffer strips and enhanced effluent treatment could decrease the load by 47 % to 18,190 kg year^{-1} in the Worka catchment (Figure 5.2.c). In the Upper Worka sub-catchments, the buffer strip scenario estimated indicates a decrease of TP loads of 45 % from 22,170 to 12,190 kg year^{-1}).

Total phosphorus

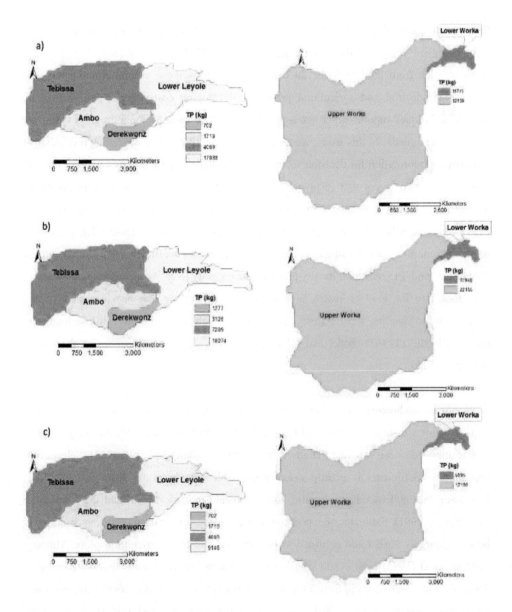

Figure 5.2. Total phosphorus (TP) loads (kg yr−1) for the scenarios of: (a) implementing riparian buffer strips, (b) enhancing of effluent treatment plants, and (c) the joint scenario of a and b in the Tebissa, Ambo, and Derekwonz sub-catchments and in the Lower Leyole River catchments and the Upper Worka sub-catchment and Lower Worka River catchment.

5.4 Discussion

When data on nutrients in soils and water are limited, or sometimes non-existent and catchments are ungauged, there is a particular need to find applicable methods to estimate nutrient loads from land to water. In catchments such as those around Kombolcha, with increasing industrial and agricultural pressures, developing cost-effective but sufficiently reliable techniques support water management as countries like Ethiopia progress with their development agenda. In this study, we evaluated a range of possible modelling methods that can provide information for decision support to both manage and build awareness of nutrient emissions to two rivers that receive both nutrient loads from agriculture and additional downstream loads from a range of industries. The screening of available models identified the PLOAD model, with its relative simplicity and few data requirements (Table 5.1.), as the most promising one to use in this situation. Although using export coefficients to estimate nutrient loads from land, PLOAD additionally provides the capacity to easily incorporate point-source emissions into the load estimates as well as a user interface that provides clear spatial visualization of loading and land use. The BMPs, which serve to reduce diffuse loads, are included in the PLOAD model and offer options to evaluate the management alternatives (USEPA, 2015).

In ungauged catchments and/or those without very frequent, even daily or less, estimates of nutrient loads are subject to potential high error. Nevertheless, reasonable accuracy to guide management has been reported (Thodsen et al., 2009; Strömqvist et al., 2012). Selection of appropriate coefficients is greatly aided from published (including from so-called grey literature) data and based on climatic, topographical, geological and land use similarities (Irvine et al., 2001). Estimates of export coefficients from land use in sub-Saharan countries is, however, largely absent and confined to loss of nutrients from agricultural plots. Although Scheren, et al (1995) used export coefficients to estimate nutrient loads from catchments in Tanzania, these were derived from a much earlier compilation of export coefficients estimated for various locations in the United States and Europe by Loehr et al. (1989). In our study, expert judgement was used to select coefficients based on similarities that would best apply to the terrain, soil and climate of Kombolcha.

Our results indicate that the model error often increases with catchment size. While optimizing the export coefficients in the calibration process, the lowest and highest export coefficient values were limited to the values from literatures (Table 5.3.). The calibrated PLOAD model estimated TN loads with 89% of reduced sum of errors compared with the TN loads in the pre-calibration estimates of the model (Table 5.5.a, c). The individual catchment errors were generally greater in the catchments than their corresponding sub-catchments (Table 5.5.c), varying from 35%, which is considered a reasonable estimation (Donigian, 2002), to 258%. Though these errors were considerably reduced compared with the PLOAD estimations of the pre-calibration, they remain large. For two of the catchments, the model still predicts the TN loads with an error of more than 100% (Table 5.5.c). For the TP loads, the sum of errors in the calibrated PLOAD model was reduced by only 5 % compared with the pre-calibrated model errors (Table 5.6.c). However, except for the Tebissa sub-catchment, for which the model resulted in a 2% error of estimation, the individual catchment errors varied from 46 to 94%. The model errors also vary in direction among the sub-catchments for both TN and TP (Table 5.5.c and Table 5.6.c). This makes it difficult to identify priority areas for intervention. In comparison with the errors in the calibration process, the model validation for TN resulted in lower sum of errors (Table 5.5.d), but this could be consequential on higher discharge during the wet season (i.e. June to September, 2014) from which the validation data set is derived (Zinabu et al., 2018).

The model errors arise from the uncertainty in the modelling parameters. As the export coefficient values are uniform for each land use within a catchment, the values disregard the land use proximity to hydrologic pathways and some attenuation of nutrients that may occur due to variations in runoff rates, plant cover, soil retention, and travel distance to streams. The PLOAD model operates at catchment scale, and scaling effects from the interaction between the land use and land characteristics can lead to higher variance with increased catchment size (Tables 5.5.a, 5.6.a). The monitoring data used for calibration and validation provides the other main potential source of uncertainty as the data were collected by periodic grab sampling and across two rainy seasons that differed in intensity (Zinabu et al., 2018). Estimating water flows at the catchments' outlets were done using rating curves and include the inherent uncertainties of measuring flow velocity, water depth and cross sectional areas (Harmel et al., 2006).

Despite the uncertainties that are inherent in this work, our model provides a foundation for developing simple modelling techniques in data poor regions with little or no capacity for regular monitoring. As more reliable data are available, such as land management practices (e.g. crop type and animal stocking), the model can be refined and used to guide future development of export coefficients applicable to semi-arid regions.

Most of the factories in Kombolcha have poor waste water treatment facilities and discharge effluents to the Leyole and Worka rivers. The effluents from the steel processing and brewery were discharged with no treatment, while the effluents from the textile, tannery and meat processing were poorly treated with old treatment facilities. Despite the presence of decades old factories, the pollution control process in the city is still at an early stage. This is coupled with poor land management and unrestricted fertilizer use, resulting in increased connectivity of the land and drainage networks and rapid transfer of pollutants (Tucker and Bras, 1998).

Our study shows that introducing riparian buffer strips and enhancing the factories effluent treatment could considerably reduce the nitrogen and phosphorus loads (Figure 5.1. and Figure 5. 2.). In the Leyole catchment, enhancing the efficiency of the effluent treatments of the four factories by 50 % could reduce both the TN and TP loads emission by 27% according to the scenario analysis.

For the scenario of riparian buffer strips, we used a load reduction efficiency of 40 and 45% for TN and TP, respectively (Horst et al., 2008). Our study indicates that riparian buffer strips of 4 m could considerably reduce nutrient loads from the catchments (Figure 5.1. and Figure 5. 2.). However, additional studies are clearly needed on effects of buffer strips. Wenger (1999) suggested that a grass or a forested buffer, with a width of between 10 and 20 meter, is needed to effectively reduce loads of nitrogen (by 50–100 %) and phosphorus (68 – 95%). Narrow width strips, for e.g. 4 meter were reported by Blanco-Canqui et al. (2004) to reduce approximately 71 % of TN and TP.

We found the PLOAD model has good potential to be a cost-effective support for management (Zhenyao et al., 2011). The graphical and tabular outputs from the PLOAD model in the BASINS system provides good visualization of relationships of pollutant sources in catchments

(USEPA, 2015). Despite the uncertainties in the parameterization, the model provides a good first-stage approach to identify where management interventions may be most effective.

5.5 Conclusions

For data-poor regions, as illustrated by the Kombolcha catchments, the PLOAD/BASINS appears as a useful model for estimating nutrient loads and evaluating management measures in a catchment. In the absence of local data for choosing effective nutrient export coefficients parameters, export coefficients from other similar catchments provides reasonable estimates of TN and TP loads in small catchments. Our result indicate that model error often increases with catchment size. Based on our study, the PLOAD model suggests a combination of improved industrial effluent treatment and riparian buffer strips in the catchments could substantially reduce TN and TP loads. With careful calibration and validation, the PLOAD model can serve an important role in planning industrial and agricultural development in data-poor nations. It is clear however that as catchment and industrial pollution increases in sub-Sahara Africa in concert with development aims, basic and detailed spatial and temporal data collection is a priority for better calibration of models and support of water quality objectives.

Chapter 6

Synthesis and conclusions

6.1 Overview

This Thesis is a contribution to river protection and better understanding of water quality management of sub-Saharan African tropical rivers and sediments. The rivers of the study area have seasonally low hydrological flows and receive effluents from several factories. Rainfed agriculture is commonly practiced in the rivers catchments and is restricted to wet seasons during which the majority of annual diffuse loads is transported into the rivers. This study is a first and small-scale monitoring with limited frequency in the wet season of two monitoring years (Chapter 2).The overall objective of the Thesis has been to quantify and evaluate the loads and transfer of four heavy metals (Cr, Cu, Zn and Pb), and nutrients (N and P) into the rivers of semi-arid catchments in north-central Ethiopia, and review related policy controls at a wider perspective of sub-Saharan Africa. This general objectives has been divided into specific objectives, as explained in Chapter 1 (section 1.3). Each Chapter starting from 2 to 5 explained specific objectives and answered particular research questions related to these objectives, and in combination, these Chapters contribute to the above overall objective. The intention of this final Chapter is to integrate the outcomes of the separate Chapters and discuss policy implications of heavy metals and nutrients loads and transfer into the rivers.

The characteristics of the study area are described in Chapter 1. This study has identified that the heavy metals and nutrients loads from the manufacturing industries of Kombolcha city is affecting the ecological health of receiving rivers. Although cost and logistic issue restricted sustained monitoring and limited the data base for effluents, the study has included monitoring of the heavy metals within the effluent mixing zones of the rivers, which is often not done in developing countries. This has improved the database for effluents and understanding of effluent loadings effect in the rivers (Chapter 2). Effluent management is found a key improvement area in the factories (Chapter 2) and weak operational capacity of the local and regional environmental institutions is major factor in prohibiting effective regulation of emission from industries (Chapter 3) Moreover, the study identifies the need to check commitment of foreign investors to environmentally sustainable industrial development policy of the county. A foreign company in the study area has been operating for a number of years without effluent treatment facilities, albeit having a high awareness to environmental protection (Chapter 2). The findings of the research contribute to the growing base of knowledge that shows an increasing trend of heavy metals and nutrient loads in rivers of industrializing regions of Ethiopia (Derso et al., 2017; Akele et al., 2016), and fills gaps in understanding of impacts and policy implication of heavy metals and nutrient pollution in the rivers of the sub-Saharan countries.

This study also explored suitable model to estimate and TN and TP loads and evaluate managements in the catchment using specific criteria and present the use of PLOAD/BASINS (Chapter 5). With no local data for choosing effective nutrient export coefficients parameters, export coefficients from other similar catchments were used in the model and provides reasonable estimates of TN and TP loads in small catchments. Poor land management plays a major role in relatively high transport of diffuse nutrient transfer into the Leyole and Worka rivers (Chapter 4). Decision support systems to manage the rivers' catchments are barely implemented, as related scientific information is scant and access to proprietary software used to estimate pollutant loads is limited (Chapter 5). The findings of this study show that transfer, loads and concentrations of heavy metals and nutrients were high and affected the quality of the Leyole and Worka rivers and sediments in the industrializing Kombolcha catchments (Chapter 3 and 4), and clearly have implications on the rivers of Ethiopia. In addition to the provision of possible direction for future research, this Thesis has provided key gaps in water quality protection measures, implementation process, enforcement and related information needs of the sub-Saharan countries from Ethiopian perspective.

6.2 Synthesis

6.2.1 Influence of industrialization on the transfer of heavy metals and nutrients in rivers and sediments

As explained in Chapter 2 and 3, industrial pollution in the Kombolcha catchments have led to accumulation of heavy metals in the receiving rivers. Heavy metals emission, especially Cr from the tannery and Zn from the steel processing factory effluents were high (Chapter 2) and have exceeded the national emission limit. As a consequence, heavy metal concentrations of the Leyole River were often found at toxic levels, both in river water and sediments (Chapter 3).

This study has shown that nutrients emissions by a local meat factory and brewery had a negative impact on the ecological health of the receiving river water. The TN and TP concentrations in these effluents exceeded the national emission guidelines. Nutrient concentrations in brewery effluents were found to be higher compared with data from other African countries (Abimbola et al., 2015; Inyang et al., 2012; Parawiraa et al., 2005) (Chapter 4). The relative concentration of nutrients in these effluents will become more important under reduced dilution of low flows of the rivers (Halling-Sörensen and Jörgensen, 2008). This Thesis exemplifies the challenges of industrialization in the city of Kombolcha and indicates that proper regulation is urgently needed to prevent further increases in nutrient and heavy metal emissions.

6.2.2 Effect of land use intensification on nutrients loads

The effect of land use on the transfer of nutrient into the Leyole and Worka rivers are discussed in Chapter 4. With no available information on the distribution and proportion of land uses in the Kombolcha catchments, this study identified seven land uses and presented their proportions in the catchments (Figure 4.2.). Intensive cropping is the largest land use practice in every of the sub-catchments compared with other land uses. The nutrient concentrations varied in the upstream but increased in the downstream along the main rivers of the catchments. Higher TN and TP transfer was observed in the sub-catchments with higher proportion of crop lands. More TP transfer were notable from the forested land in a hilly landscapes of the uplands. The study also shows that more particulate phosphorus are associated with prevalence of land degradation in the form of soil erosion in the catchments, as more particulate P was transferred compared with dissolved P into the rivers (Chapter 4). This problem will clearly exacerbate with increasing population and agriculture encroachment and cultivation onto the hilly landscape of the catchments (Gashaw et al., 2014). Using source apportionments, estimates of TN and TP loads from lands were found much higher compared with the factory units and this highlights the importance of diffuse nutrient loads in the Kombolcha catchments (Chapter 4).

6.2.3 Concerns for industrial effluents and lands in the Kombolcha catchments

Chapter 3 illustrates that each factory in the Leyole and Worka river catchments is managed independently. Despite the close proximity of the factories and disposal of the effluents into the two nearby rivers, little effort has been done to collectively manage pollutants in the effluents. At the time of this study, no treatment facilities were present for the brewery and steel processing factory and the other three factories use lagoons or retaining ponds to treat their respective effluents. However, these facilities are quite old and designed to treat organic and sediment wastes but not dissolved heavy metals and nutrients. These factories are required to comply with national emission limits for heavy metals and nutrients, but inspections are not done by the local or regional environmental protection institutions (Afework et al., 2010; EEPA, 2010; FDRE, 2002b). Such uncontrolled waste emission are common in most industrial parts of Ethiopia (Derso et al., 2017; Akele et al., 2016; Larissa et al., 2013), and this Thesis presents suggestions for improvement in the monitoring and control of industrial effluent, and possible limitations of correction measures.

Kombolcha's catchments comprise relatively small sub-catchments with steep and flat landforms in a semi-arid agro-climate and low hydrological flows of rivers. Poor land managements like soil erosion are evident in the croplands, which are often dissected by gulley, and grazing lands in the hilly landscapes. Chapter 4 shows that there is quite high transfer of TP and prevalence of land degradation in these lands. Chapter 4 also clarifies that high percentage of crop lands in the sub-catchments is associated with increasing TN transfer in the rivers. In the croplands, more fertilizers are used to cope with the demand for intensive crop production, and no legal restriction is in place for the fertilizer application rates. The application is considerably higher compared with available data from European countries (Velthof et al., 2014). This research suggests that both the natural landforms and mismanagements on lands have contributed to the transfer and accumulation of more nutrients at downstream (Soranno et al., 2015; Duncan, 2014; Gasparini et al., 2010). Furthermore, Chapter 4 provides useful information on potential contribution of human inputs from the commonly practiced open defecation in the Kombolcha catchment. Interestingly, the export of TN from open defecation was considerably high compared to loads of the factory effluents. Given that Ethiopia has one of the highest number of people openly defecating in the World (WHO/UNICEF, 2014), this research is first step towards enhancing the need to assess human impacts not only for nutrient loads but also for public health.

6.2.4 Effects of heavy metals and nutrients loads in quality of river and sediments

Chapter 3 and 4 emphasize on the impacts of heavy metal and nutrient pollutions in the Leyole and Worka rivers. The pollutions were evaluated using environmental quality guidelines for heavy metal concentrations in water, compiled by Macdonald et al. (2000). Findings of this research show concentrations of Cr, Cu and Zn surpassed the guidelines for aquatic life, human water supply, and irrigation and livestock water supply For sediment quality, the numerical Sediment Quality Guidelines (SQGs) were used to understand potential effects of the heavy metals on aquatic lives in the rivers (MacDonald et al., 2000b; USEPA, 1997a). All heavy metals exceeded guidelines for sediment quality for aquatic organisms (Chapter 3). A normalization process for sediment sizes was applied based on organic matter and grain size distributions and thus more toxicity effect of the heavy metals in the sediments was revealed for the rivers (Akele et al., 2016; Department of Soil Protection, 1994). This study shows that the normalization is useful in overcoming both texturally and organic matter driven variations of toxic concentrations in sediments of rivers.

The effects of the nutrients loads into the rivers are also described in Chapter 4. The TN and TP concentrations in the downstream section of the Leyole and Worka rivers had exceeded the commonly used quality standards, especially for irrigation and livestock water supply. With the findings of higher nutrients downstream compared with the rivers' upstream, this Thesis shows that the clear influence of land uses and factory emissions on the downstream water quality. The study also highlights the problem with industrialization policy of Ethiopia and many parts of Africa (Chapter 2), and demonstrates the importance of building knowledge and capacity for better monitoring and management of rivers and other water bodies (Chapter 4).

6.3 Implications of the study

This Thesis has verified that the transfer and loads of heavy metals and nutrients into the Leyole and Worka rivers is a serious environmental issue in Kombolcha. Many of the Kombolcha factories were found to have old and ineffective effluent treatment facilities. Like most sub-

Saharan countries, monitoring resources and operational capacity of local institution are weak (Chapters 3 and 4). This research provides evidence that the current industrial parks envisaged for Kombolcha city will present further major environmental risks to the rivers of Kombolcha. The situation in the city exemplifies heavy metals pollution across Ethiopia (Derso et al., 2017; Akele et al., 2016). With the Ethiopian government plan for more number of industrial parks across the country, this research promotes that effective effluent management and strong regulatory structures aided by monitoring of effluent receiving water bodies are key improvement areas to achieve long-term development. In order to achieve these, appropriate and locally supported pollution control measures are urgently needed. Furthermore, the Thesis emphasis that it is necessary to facilitate the local environmental controlling institutions with the required instrumentations and mechanisms for law enforcement.

Industrial effluent management need

As explained in Chapter 2, many industrial technologies in Kombolcha are quite old and there is a tendency to import cheaper technologies to cope with environmental requirements under increasing pressure of economical returns. A number of studies have reported that with the absence of government initiatives to finance cleaner production, waste treatment facilities is a high burden for investors in Ethiopia (CEPG, 2012; EEPA, 2010; Getu, 2009). Given the government strategy of expansion of industrial parks in specific zones across the regions of the country, this research suggests the idea of efficient and cost effective initiative measures to address the environmental concern of effluents in the zones. Integrating a single centralized waste treatment facility used by multiple industries within an industrial park could be useful measure for sustainable industrial development. While the Government may lease treatment facilities, the operation of the facilities can be financed in collaboration of the industry owners in the park.

The industrial effluent guidelines of Ethiopia are based on industrial categories and use Best Available Technologies (BAT) permits as precautionary measures (Chapter 2). No guidelines (for ecological protections) are set for effluent receiving waters, and this has made it impossible to exactly understand impacts of effluent emissions into the receiving waters (Chapters 3). To address the environmental risk of emissions from the industrializing parks, the governing body has to emphasize on conserving the tolerance of the surrounding environment in enduring the impacts from the emissions. This study suggests that policy makers should encourage developing emission criteria of heavy metals based on the carrying capacity of the effluent receiving rivers instead of the currently emission standards developed based on the type of the factory (EEPA, 2010).

Monitoring capacity need

In Ethiopia, provision of monitoring information is poor and impacts on rivers and sediments water quality information are often unknown (Chapters 2 and 3). Chapter 3 discusses that lack of monitoring infrastructures and poor management and scientific capacity in the institutions is hampering monitoring of rivers (Chapter 3). This research infers that developing monitoring protocols and institutional capacities are needed to support Ethiopia in its ambitions for industrialisation. Furthermore, given the poor human resources, expertise and infrastructure in monitoring, this research shows the possibility of using verifiable land and water quality models in guiding monitoring and management of nutrients in data-poor catchments. After screening several models, the PLOAD model is found cost effective, adequately estimating nutrient loads, and provide good visualization of relationships of pollutants in the Kombolcha catchments, although with the uncertainties in the parameterization (Chapter 5). This research suggests that the model can serve an important role in planning industrial and agricultural development ensuring reduced nutrient loads into rivers. The broad implication of this research is that commitment from local and national governments is necessary in developing institutional capacity and implementing low cost and adaptable monitoring techniques. In this aspect, this research has highlighted that advocacy for public and private partnership is useful to overcome limited governmental institutional structures and lack of adequate instruments and deficiencies in necessary skills in monitoring (Chapter 3).

Policy improvement

Agricultural intensification and poor land management are found as major sources of nutrients loads in Kombolcha catchments (Chapter 4). Given the industrialization policy of Ethiopia, which is based on expanding food processing, garments and beverage industries (MoFED, 2002), mainly using raw materials from the country's vast agricultural productions , these sources are key environmental issues in Ethiopia. The growing floriculture industry has substantially increased fertilizers applications, adding to current reports of high emissions

(Endale, 2011; Getu, 2009). This research underlines the necessity of regulated fertilizer application and controls on nutrient emissions from catchments, The study suggests that extension services, supported by policy framework and political commitment, designed to optimize crop production and reduce nutrient loads could benefit such services and promote sustainable agriculture and land use.

Despite Ethiopian government awareness of potential impacts from pollution, there is an obvious limited action to protect human or ecosystem health. Low levels of financing for environmental research and monitoring has hindered availability of reliable information on water quality and undermined the capacity to develop national water quality guidelines (Chapter 3). This study indicates that action for agreed standards of quality and a relevant policy framework that supports monitoring and regulation is highly needed to control emissions (Chapters 2, 3, and 4). Currently, Ethiopia is following the WHO guidelines for drinking water quality, which are not designed for monitoring ecological health. To tackle these issues, this Thesis has put forward two reasonable policy approaches (Chapter 3); 1) reviewing applicability of international or neighbouring countries policies and guidelines or 2) building a monitoring network that provides baseline data to inform national policy (as done, for example, in Ghana and Kenya).

Furthermore, Chapter 3 highlights that effective and locally relevant monitoring frameworks are important to engage citizens in science, even in remote and rural areas. On this basis, the possibility of using the growing information technology in Ethiopia, like smart phones, is proposed in transmitting monitoring data and raise local awareness on water quality (Katsriku et al., 2015; Danielsen et al., 2011). This Thesis has shown that important changes are needed in environmental institution in using these data and mainstreaming GIS and remote sensing techniques for monitoring of rivers and modelling of water quality especially in large basins (Dube et al., 2015; Ritchie et al., 2003).

6.4 Recommendations for further Research

One of the most important issues in estimating pollutant loads apportionment is the sampling of the river water and the corresponding analysis of the sampled water. Although this study provides the first estimates of heavy metal and nutrient loads in Kombolcha catchments, the sampling frequency was limited to a 15-days in eight months across two successive years, and this likely underestimated concentration of nutrients from diffuse and point sources of the catchments. Future study should consider more sustained monitoring frequency and years that include both high and low rainfall events and intermittent chemical emissions from point sources. Deployment of automated samplers using either high intensity or continues monitoring automated samplers are recommendable alternatives. Additionally, in Chapter 4, the source apportionment estimation of the diffuse sources was based on subtraction of point sources loads from total loads of a catchment. However, future studies should aim to more direct estimation from the diffuse sources, especially the land uses. Direct measurement of diffuse pollutants in surface flows of waters, either by natural or using simulated rainfalls, on selected land uses of a catchment is an option. This approach is helpful not only in estimating diffuse loads but also in assessing loads from each land use and using of such information in modelling processes.

More studies are needed to understand the influence of effluent management and treatment technologies of the factories in Kombolcha city. Comparing water quality before/after improvements in treatment technologies, and quantifying the cost of heavy metals treatment vis-a-vis monetary valuation of the damage effects of the heavy metals in receiving rivers are important issues in identifying improvement areas and decision making processes.

In a preliminary investigation to examine the first flush effects, waters were collected for the first rainfall event of the wet season (i.e. June – September, 2014) at a sub-catchment outlet in the upper part of the Leyole River catchment. The sampling analyses showed that 80% of TSS, TKN and TP masses are transported in the first 67%, 63% and 67% volume of flows respectively. According to the definitions of Bertrand et al. (1998) and Taebi and Droste (2004), the first flush effect has been met, as the TSS, TKN and TP showed normalized mass-volume curves that are diverged outward from the bisector line in the early volume of flows, which is indicating large transport of TSS, TKN and TP in the first flushes flows into the sub-catchment outlet. Additional research is needed in multiple rain events and at more number of areas, including the peri-urban and urban catchments, to define and quantity better representative first flushes TSS, TKN and TP mass and volume for the Kombolcha catchments. This will be useful in planning of surface flush water quantities needed to treat and remove TSS and nutrients, and designing of treatment facilities such as ponds, wetlands, infiltration, and filtration measures.

This Thesis estimated the loads of four heavy metals from five factory units in the Kombolcha catchments. There is a need to extend the study towards the potential emission of these heavy metals from other point or diffuse sources in the study area. Since more heavy metals, for e.g. Cd, Hg and Ni, are suspected to be released especially from the steel processing industry, future studies should consider monitoring of these heavy metals emissions from the industries as well. Furthermore, Chapter 3 brings out the presence of occasional high heavy metal concentrations in the upstream sites of both the Leyole and Worka rivers. All factory units are located downstream of these sites, and therefore, have no contribution to the high concentration of heavy metals in the sites. The city's landfills, which are unlined and open-pits, are close to these upstream sites. Solid wastes especially from the nearby factories are dumped there and additional study is required to quantify the potential surface and subsurface transfer of heavy metals into the rivers. This will be particularly helpful in evaluating waste management issues in the city of Kombolcha.

This research highlights the importance of anthropogenic factors, in particular land uses and factory units, in nutrient transfer of the Kombolcha catchments. However, Kombolcha is topographically varied with a rural upland landscape and lowland urban areas that are prone to erosion and flooding, respectively. The nutrient transfers are affected by various hydrological pathways which depend on the landscape's hydro-meteorological characteristics (Van der Perk, 2006). Given the erratic nature of rainfall in Kombolcha catchments, occasional storms on the hilly landscapes of the uplands can cause high overland flows and flooding in the lowlands' flat lands. This study provides useful information on the influence of human land use on nutrient loads. However, there is a need to identify main predictors, both from anthropogenic and natural landscape characteristics, of nutrient loads. Additional study is required to understand the influence of natural landscape features, for e.g. slope and soils, on the nutrient loads. This would particularly help to prioritize problems in decision making and design effective nutrient management for catchments.

As explained in Chapter 4, there are high TN and TP loads in the effluents of the meat processing factory and brewery. Large differences were found between concentrations of TN and the sum of $(NH_4 + NH_3)$–N and NO_3–N, and also, the TP concentrations were much higher than those of PO_4–P. This shows the presence and dominance of other pollutant forms, like organic N and P, in the effluents of these factories. Additional study is therefore needed to exactly identify and quantity these forms. Furthermore, the low PO_4-P/TP concentration proportions found in all catchments outlets, and the finding of significant difference in the concentrations of PO_4–P between the upstream sites and their corresponding catchment outlets (Chapter 4, Table 4.2.), shows the presence of other forms of phosphorus transported in the catchments' rivers. Considering the evident soil erosion problem in the area (Chapter 4), further study is required to assess whether particulate phosphorus (PP) is dominant in TP loads. This is important to reflect on the prevalence of land degradation, especially associated with high slope landforms used for cultivation and livestock grazing of the Kombolcha catchments. Additionally, this research presents an initial assessment of relationship between river sediments and heavy metal pollution, looking into the concentrations of the heavy metals in sediments based on five groups of grain sizes (Chapter 3). Highest heavy metal adsorption capacities can be expected for fine grained (< 63 μm) sediments, because of their larger specific surface area (Devesa-Rey et al. 2011; Wang 2000) (Chapter 3). However, Chapter 3 also shows the trend of decreasing heavy metal concentration with increasing grain sizes is not straightforward (see Fig. 3.4). This suggest that more study is needed to understand the effect of smaller grain sizes in order to have a clearer trend of the heavy metal concentrations in further differentiations within the <63 μm fraction.

This Thesis presents a calibrated PLOAD model that is particularly useful for characterization of nutrient pollution in the data-poor Kombolcha catchments. The error of estimation in the PLOAD model was found increasing with catchment size and have considerable uncertainties in the model parameterization. Chapter 6 discusses that these errors are related to the uncertainties in the modelling parameters. In addition to scaling effects from the interaction between the land use and land characteristics, model parameters including export coefficient values, the monitoring data that were used for calibration and validation, and water flows data of the catchments' outlets were identified as the main source of errors. The export coefficients values are crucial modelling parameter for the PLOAD model, but this study has used literature-based values due to unavailability of local and regional export coefficient data for N and P. Most of this information is available only from studies done in either European or American catchments, which are quite different in many aspect and less relevant to conditions of sub-Saharan countries. In-situ measurements of the export coefficients from representative land uses in catchments can help better estimation of N and P in future studies. Moreover, amendment is required to the other uncertainties so as to enhance the performance of the

PLOAD model in the Kombolcha catchments, and future study should involve basic and detailed spatial and temporal data generation that would minimize uncertainties. With further careful calibration and validation, the PLOAD model can then serve an important role in planning industrial and agricultural development.

This study has shown that the Leyole and Worka rivers contribute high loads of heavy metals and nutrients to the Borkena River. The Borkena River replenishes the "Cheffa wetland", which is located at closer distance of downstream Kombolcha city. The Cheffa wetland (also called the "Borkena Valley") is about 82,000 ha (Tamene, et al., 2000), and vitally supports large numbers of pastoralists and their cattle from both the Afar and Oromia regional states of Ethiopia (Piguet, 2002). Being one of the richest diverse aquatic plants and animals wetlands of Ethiopia, the Cheffa wetland contributes to making the surround areas liveable (Getachewa, et al., 2012). With agricultural intensification, encroachment into the wetland, and growing industrialization in the city of Kombolcha, heavy metals and nutrient loads into the wetland is an environmental issue for future research to explore impacts on ecology of the Cheffa wetlands.

References

Abbaspour, S. 2011. Water quality in developing countries, south Asia, south Africa, water quality management and activities that cause water pollution. In: International conference on environmental and agricultural engineering. IACSIT Press, Singapore.

Abdel-Satar, A., Ali, M. & Goher, M. 2017. Indices of water quality and metal pollution of Nile River, Egypt. Egyptian Journal of Aquatic Research 43: 21-29.

Abimbola, M., Josiah, A., Sheena, K., Feroz, M. & Faizal, B. 2015. Characterization of brewery wastewater composition. International Journal of Environmental, Chemical, Ecological, Geological and Geophysical Engineering 9: 1073-1076.

Aceves-Bueno, E., Adeleye, A., Bradley, D., Tyler Brandt, W., Callery, P., Feraud, M., Garner, K., Gentry, R., Huang, Y. & McCullough, I. 2015. Citizen science as an approach for overcoming insufficient monitoring and inadequate stakeholder buy-in in adaptive management: criteria and evidence. Ecosystems 18: 493-506.

Adakole, J. A. & Abolude, D. S. 2009. Studies on effluent characteristics of a metal finishing company, Zaria, Nigeria. Journal of Environmental and Earth Sciences 1: 54-57.

Ademe, A. S. & Alemayehu, M. 2014. Source and determinants of water pollution in Ethiopia: distributed lag modeling approach. Intellectual Property Rights 2: 110-114.

Afework, H., Alebachew, A., Demel, T., Habtemariam, A., Meskir, T., Terefe, D. & Wondwossen, S. 2010. Ethiopian environment review. Eclipse Printing Press, Addis Ababa.

African Development Bank. 2017. Gender, poverty and environmental indicators on African countries. Scanprint, Horsens.

African Union 2018. Annual report on the activities of the african union and itsorgans African Union, Addis Ababa.

Afum, B. & Owusu, C. 2016. Heavy Metal Pollution in the Birim River of Ghana. International Journal of Environmental Monitoring and Analysis 4: 65-74.

Ahmed, G., Miah, M. A., Ahmad, J. U., Chowdhury, D. A. & Anawar, H. M. 2012. Influence of multi-industrial activities on trace metal contamination: an approach towards surface water body in the vicinity of Dhaka Export Processing Zone. Environmental Monitoring and Assessment 184: 4181-4190.

120

Akan, J. C., Moses, E. A., Ogugbuaja, V. O. & Abah, I. 2007. Assessment of tannery industrial effluents from Kano metropolis, Kano State, Nigeria. Journal of Applied Sciences 7: 2788-2793.

Akele, M. L., Kelderman, P., Koning, C. W. & Irvine, K. 2016. Trace metal distributions in the sediments of the Little Akaki River, Addis Ababa, Ethiopia. Environmental Monitoring and Assessment 188: 389-402.

Aklilu, A. 2013. Heavy metals concentration in tannery effluents associated surface water and soils at Ejersa area of East Shoa, Ethiopia. Journal of Environmental Science and Toxicology 1: 156-163.

Alberta Environmental Protection 1993. Alberta ambient surface water quality interim guidelines. Alberta Environmental Protection, Environmental Assessment Division, Edmonton.

Alcamo, J., Fernandez, N., Leonard, S., Peduzzi, P., Singh, A. & Harding Rohr Reis, R. 2012. 21 issues for the 21st century: results of the unep foresight process on emerging environmental issues. UNEP, Nairobi.

Alexa, V. 2013. Issues for monitoring the pollutants in wastewaters and the environmental management system in metallurgy. Journal of Environmental Protection and Ecology 14: 618-628.

Allan, J. D. 2004. Landscapes and riverscapes: The influence of land use on stream ecosystems. Annual Review Of Ecology, Evolution, and Systematics 35: 257-284.

Allison, J. D. & Allison, T. L. 2005. Partition coefficients for metals in surface water, soil, and waste U.S. Environmental Protection Agency Office of Research and Development Washington D.C.

Alonso, R. B., Andrés, G. G. & César, Á. D. C. 2016. Definition of mixing zones in rivers. Environmental Fluid Mechanics 16: 209-244.

Álvarez-Romero, J. G., Wilkinson, S. N., Pressey, R. L., Ban, N. C., Kool, J. & Brodie, J. 2014. Modeling catchment nutrients and sediment loads to inform regional management of water quality in coastal-marine ecosystems: A comparison of two approaches. Journal of Environmental Management 146: 164-178.

Ambrose, R. B., Wool, T. A., Connolly, J. P. & Schanz, R. W. 1988. WASP4, a hydrodynamic and water-quality model-model theory, user's manual, and programmer's guide. US Environmental Protection Agency, Athens.

Anderson, C. W. & Rounds, S. A. 2010. Use of continuous monitors and autosamplers to predict unmeasured water-quality constituents in tributaries of the Tualatin River, Oregon. USGS, Oregon.

Anderson, J. 1976. A land use and land cover classification system for use with remote sensor data. US Government Printing Office, Washington D.C.

Arimoro, F. O., Ikomi, R. B. & Iwegbue, C. M. 2007. Water quality changes in relation to Diptera community patterns and diversity measured at an organic effluent impacted stream in the Niger Delta, Nigeria. Ecological Indicators 7: 541-552.

Armitage, P. D., Bowes, M. J. & Vincent, H. M. 2007. Long-term changes in macroinvertebrate communities of a heavy metal polluted stream: the River Nent (Cumbria, UK) after 28 years. River Research and Applications 23: 997–1015.

Arnold, J., Srinivasan, R., Muttiah, R. & Williams, J. 1998. Large area hydrologic modeling and assessment part I: model development. Journal of the American Water Resources Association 34: 73-89.

Aschale, M., Sileshi, Y., Kelly-Quinn, M. & Hailu, D. 2016. Evaluation of potentially toxic element pollution in the benthic sediments of the water bodies of the city of Addis Ababa, Ethiopia. Journal of Environmental Chemical Engineering 4: 4173-4183.

Assefa, T. 2008. Digest of ethiopia's national policies, strategies and programs. Forum for Social Studies, Addis Ababa, Ethiopia.

Awoke, A., Beyene, A., Kloos, H., Goethals, P. & Triest, L. 2016. River water pollution status and water policy scenario in Ethiopia: Raising awareness for better implementation in developing countries. Environmental management 58: 694–706.

Awulachew, S. B., Erkossa, T. & Namara, R. E. 2010. Irrigation potential in Ethiopia: Constraints and opportunities for enhancing the system. IWMI, Addis Ababa.

Awulachew, S. B., Yilma, A. D., Loulseged, M., Loiskandl, W., Ayana, M. & Alamirew, T. 2007. Water resources and irrigation development in Ethiopia. IWMI, Addis Ababa.

Ayalew, W. & Assefa, W. 2014. Bahir Dar tannery effluent characterization and its impact on the head of Blue Nile River. African Journal of Environmental Science and Technology 8: 312-318.

Bartley, R., Speirs, W. J., Ellis, T. W. & Waters, D. K. 2012. A review of sediment and nutrient concentration data from Australia for use in catchment water quality models. Marine Pollution Bulletin 65: 101-117.

Bechtold, J. S., Edwards, R. T. & Naiman, R. J. 2003. Biotic versus hydrologic control over seasonal nitrate leaching in a floodplain forest. Biogeochemistry 63: 53-72.

Bertinelli, L., Strobl, E. & Zou, B. 2006. Polluting technologies and sustainable economic development. International Journal of Global Environmental Issues 10: 1-29.

Bertinelli, L., Strobl, E. & Zou, B. 2012. Sustainable economic development and the environment: theory and evidence. Energy Economics 34: 1105-1114.

Bertrand, J. L., Chebbo, G. & Saget, A. 1998. Distribution of Pollutant Mass vs Volume in Stormwater Discharges and the First Flush Phenomenon. Water research. 32: 2341

Besser, J. M., Allert, A. L., Hardesty, D., Ingersoll, C. G., May, T. W., Wang, N. & Leib, K. J. 2001. Evaluation of metal toxicity in streams of the upper Animas River watershed. U.S. Geological Society Biological Science, Colorado.

Beyene, A., Addis, T., Kifle, D., Legesse, W., Kloos, H. & Triest, L. 2009a. Comparative study of diatoms and macroinvertebrates as indicators of severe water pollution: case study of the Kebena and Akaki rivers in Addis Ababa, Ethiopia. Ecological Indicators 9: 381-392.

Beyene, A., Legesse, W., Triest, L. & Kloos, H. 2009b. Urban impact on ecological integrity of nearby rivers in developing countries: the Borkena River in highland Ethiopia. Environmental Monitoring & Assessment 153: 461-476.

Bicknell, B., Imhoff, J., Kittle, J., A., D. & Johanson, R. 1993. Hydrologic Simulation Program—FORTRAN (HSPF): User's Manual for Release 10. US EPA Environmental Research Lab, Athens.

Blanco-Canqui, H., Gantzer, C., Anderson, S., Alberts, E. & Thompson, A. 2004. Grass barrier and vegetative filter strip effectiveness in reducing runoff, sediment, nitrogen, and phosphorus loss. Soil Science Society of America Journal 68: 1670-1678.

Boggs, S. 2009. Petrology of sedimentary rocks. Cambridge University Press, Cambridge.

Bouwman, A. F., Van Drecht, G., Knoop, J. M., Beusen, A. H. & Meinardi, C. R. 2005. Exploring changes in river nitrogen export to the world's oceans. Global Biogeochemical Cycles 19: 1002-1016.

Bowden, K. & Brown, S. R. 1984. Relating effluent control parameters to river quality objectives using a generalised catchment simulation model. Water Science and Technology. 16: 197–205.

Bowes, M. J., Smith, J. T., Jarvie, H. P. & Neal, C. 2008. Modelling of phosphorus inputs to rivers from diffuse and point sources. Science of the Total Environment 395: 125-138.

Brack, W. 2015. The solution project: challenges and responses for present and future emerging pollutants in land and water resources management. Science of the Total Environment 503: 22-31.

Brown, L. C., . & Barnwell, T. O. 1987. The enhanced stream water quality models QUAL2E and QUAL2E-UNCAS: documentation and user manual. US EPA, Georgia.

Carpenter, S. R., Caraco, N. F., Correli, D. L., Howarth, R. W., Sharpley, A. N. & Smith, V. H. 1998. Nonpoint pollutionof surface waters with phosphorus and nitrogen. Ecological Applications 8: 559-568.

CCREM (Canadian Council of Ministers of the Environment) 2001. Canadian water quality guidelines. Environmental Quality Guidelines Division, Ottawa.

CEPG (Centre for Environmental Policy and Governance) 2012. Environmental policy update 2012: development strategies and environmental policy in East Africa. Colby College Environmental Studies Program Waterville, Maine.

Chapman, D. V. 1996. Water quality assessments : a guide to the use of biota, sediments, and water in environmental monitoring. E and FN Spon, London.

Chiew, F. & McMahon, T. 1999. Modelling runoff and diffuse pollution loads in urban areas. Water Science and Technology 39: 241-248.

Chikanda, A. 2009. Environmental degradation in sub-Saharan Africa. In: Environment and health in sub-Saharan Africa: managing an emergency crisis, Luginaah, I., Yanful, E. (Eds). Springer, Dordrecht.

Choudhury, A. 2006. Textile preparation and dyeing. Science publishers, New Hampshire.

Commission, E. 2011. Our life insurance, our natural capital: an EU biodiversity strategy to 2020. European Commission, Brussels.

Corcoran, E., Nellemann, C., Baker, E., Bos, R., Osborn, D. & Savelli, H. 2010. Sick water? The central role of wastewater management in sustainable development: a rapid response assessment. Birkeland Trykkeri AS, Birkeland.

Crutzen, P. & Steffen, W. 2003. How long have we been in the anthropocene era? An editorial comment. Climatic Change 61: 251-257.

Damtie, M. & Bayou, M. 2008. Overview of environmental impact assessment in Ethiopia: gaps and challenges. MELCA Mahiber, Addis Ababa.

Dan'azumi, S. & Bichi, M. 2010. Industrial pollution and heavy metals profile of challawa river in Kano, Nigeria. Journal of Applied Sciences in Environmental Sanitation 5: 23-29.

Danielsen, F., Skutsch, M., Burgess, N., J., M. P., Andrianandrasana, H., Karky, B., Lewis, R., Jon, C., Massao, J., Ngaga, Y., Phartiyal, P., Køie, M., Singh, S., Solis, S., Sørensen, M., Tewari, A., Young, R. & Zahabu, E. 2011. At the heart of REDD+: a role for local people in monitoring forests? Conservation Letters 4: 158-167.

Darghouth, S., Ward, C., Gambarelli, G., Styger, E. & Roux, J. 2008. Watershed mananagement approaches, policies, and operations: Lesson for scaling up. World Bank, Washington D.C.

Das, J. 2014. Analysis of river flow data to develop stage-discharge relationship. International Journal of Research in Engineering and Technology 3: 76-80.

Daughton, C. G. 2014. The Matthew Effect and widely prescribed pharmaceuticals lacking environmental monitoring: Case study of an exposure-assessment vulnerability. Science of the Total Environment 466: 315-325.

Deepali, K. K. 2010. Metals concentration in textile and tannery effluents, associated soils and ground water New York Science Journal 3: 82-89.

Degens, B. P. & Donohue, R. D. 2002. Sampling mass loads in rivers: a review of approaches for identifying, evaluating and minimising estimation errors. Water and Rivers Commission, East Perth.

Delkash, M., Al-Faraj, F. & Scholz, M. 2018. Impacts of anthropogenic land use changes on nutrient concentrations in surface waterbodies: A review. CLEAN Soil Air Water 46: 51-61.

Demeke, Y. & Aklilu, N. 2008. Alarm bell for biofuel development in Ethiopia: the case of Babille elephant sanctuary. In: Agrofuel development in Ethiopia: Rhetoric, reality and recommendations, Tibebwa, H., Negusu, A. (Eds). Forum for Environment, Addis Ababa.

Department of Soil Protection. 1994. The Netherlands intervention values for soil remediation. In: The Netherlands soil contamination guidelines. Dutch Ministry of Infrastructure and the Environment, Utrecht.

Derso, S., Kidane, A., Tesfaye, K., Gizaw, M., Abera, D., Getachew, M., Abate, M., Beyene, Y., Assefa, T. & Assefa, Z. 2017. Pollution status of akaki river and its contamination effect on surrounding environment and agricultural products Ethiopian Minstry of Health, Addis Ababa.

Devesa-Rey, R., Díaz-Fierros, F. & Barral, M. 2011. Assessment of enrichment factors and grain size influence on the metal distribution in riverbed sediments (Anllons River, NW Spain). Environmental Monitoring and Assessment 179: 371-388.

De Wit, M. J. 1999. Nutrients fluxes in the Rhine and Elbe basins. Nederlandse Geografische Studies 259, Utrecht.

DHI. 1998. MIKE 11: a microcomputer based modeling system for rivers and channels. Reference manual of the Danish Hydraulic Institute, Hoersholm.

Ding, X., Shen, Z., Hong, Q., Yang, Z., Wu, X. & Liu, R. 2010. Development and test of the export coefficient model in the upper reach of the Yangtze River. Journal of Hydrology 383: 233-244.

Donigian, A. S. 2002. Watershed model calibration and validation: The HSPF experience. Water Environment Federation, Oxford.

Dorioz, J. M., Wang, D., Poulenard, J. & Trevisan, D. 2006. The effect of grass buffer strips on phosphorus dynamics-A critical review and synthesis as a basis for application in agricultural landscapes in France. Agriculture, Ecosystems and Environment 117: 4-21.

Driscoll, C. T., Whitall, D., Aber, J., Boyer, E., Castro, M., Cronan, C., Goodale, C. L., Groffman, P., Hopkinson, C. & Lambert, K. 2003. Nitrogen pollution in the northeastern United States: sources, effects, and management options. BioScience 53: 357-374.

Dube, T., Mutanga, O., Seutloali, K., Adelabu, S. & Shoko, C. 2015. Water quality monitoring in sub-Saharan African lakes: a review of remote sensing applications,. African Journal of Aquatic Science 40: 1-7.

Duffus, J. H. 2002. "Heavy metals"-a meaningless term? Pure Appllied Chemistry 74: 793-807.

Duncan, R. 2014. Regulating agricultural land use to manage water quality: the challenges for science and policy in enforcing limits on non-point source pollution in New Zealand. Land Use Policy 41: 378-387.

Dwina, R., Pertiwi, A. & Anindrya, N. 2010. Heavy metals (Cu and Cr) pollution from textile industry in surface water and sediment: a case of Cikijing river, West Java, Indonesia. In: The 8th International Symposium on Southeast Asian Water Environment, Hiroaki, F. (ed), Phuket, Thailand.

Dybas, C. L. 2005. Dead zones spreading in world oceans. BioScience 55: 552-557.

Economist Intelligence Unit. 2008. Country Report, Ethiopia Economist intelligence unit limited, London.

Edwards, C. & Miller, M. 2001. PLOAD Version 3.0 User's Manual. USEPA, Washington D.C.

EEPA (Ethiopian Environmental Protection Authority) 2002. Environmental impact assessment proclamation. Negarit Gazeta, Addis Ababa.

EEPA. 2010. Environmental management programme of the plan for accelerated sustainable development to eradicate poverty 2011-2015 Ethiopian Environmental Protection Authority, Addis Ababa.

El-Bouraie, M., El-Barbary, A., Yehia, M. & Motawea, E. 2010. Heavy metal concentrations in surface river water and bed sediments at Nile Delta in Egypt. Journal of Suo-Mires and peat 61: 1-12.

Emmanuel, B. & Adepeju, O. 2015. Evaluation of tannery effluent content in Kano metropolis, Kano State Nigeria. International Journal of Physical Sciences 10: 306-310.

EMoI (Ethiopian Ministry of Industry) 2014. Environmental and social management framework for Bole Lemi and Kilinto industrial zones competitiveness and job creation project. Ethiopia Ministry of Industry, FDRE, Addis Ababa.

EMoWIE (Ethiopian Ministry of Water Irrigation and Energy) 2016. Existing water quality situation in Ethiopia. Ministry of Water, Irrigation and Electricity, FDRE, Addis Ababa.

EMoWR (Ethiopian Ministry of Water Resources) 2004a. Ethiopian water resource management regulation. Federal Nagarit Gazeta, Addis Ababa.

EMoWR. 2004b. National Water Development Report for Ethiopia FDRE Minstry of Water and Resources, Addis Ababa.

Endale, K. 2011. Fertilizer consumption and agricultural productivity in Ethiopia. EDRI, Addis Ababa.

Enderlein, U. S., Enderlein, R. E. & Williams, W. P. 1997. Water Quality Requirements. In: Water pollution control: a guide to the use of water quality management principles, Helmer, R., Hespanhol, I. (Eds). WHO/UNEP London.

EPA. 2001. PLOAD Version 3.0 an arcview gis tool to calculate nonpoint sources of pollution in watershed and stormwater projects: User's manual. EPA, Washington D.C.

EPA. 2017. The problem: Nutrient pollution. EPA, Washington D.C.

Erni, M., Drechsel, P., Bader, H., Scheidegger, R., Zurbruegg, C. & Kipfer, R. 2010. Bad for the environment, good for the farmer? Urban sanitation and nutrient flows. Irrigation and drainage systems 24: 113-125.

ESRI. 2011. ArcGIS Desktop: Release 10. Environmental Systems Research Institute, Redlands.

Fauvel, B., Cauchie, H., Gantzer, C. & Ogorzaly, L. 2016. Contribution of hydrological data to the understanding of the spatio-temporal dynamics of F-specific RNA bacteriophages in river water during rainfall-runoff events. Water Resources 94: 328-340.

FDRE. 2002a. Environmental pollution control proclamation Federal Negarit Gazeta, Addis Ababa.

FDRE. 2002b. Environmental protection organs establishment proclamation. Federal Negarit Gazeta, Addis Ababa.

FDRE. 2016. Growth and Transformation Plan II (GTP II). National Planning Commission, Addis Ababa.

Floqi, T., Vezi, D. & Malollari, I. 2007. Identification and evaluation of water pollution from Albanian tanneries. Desalination 213: 56-64.

Fox, J. & Weisberg, S. 2011. Functions and Datasets to Accompany. In: An R Companion to Applied Regression. Sage, Los Angeles.

Francis, C. F. & Lowe, A. T. 2015. Application of strategic environmental assessment to the Rift Valley Lakes Basin master plan,. In: Monitoring and modelling dynamic environments, Dykes, A. P., Mulligan, M., Wainwright, J. (Eds). John Wiley & Sons, Ltd, Chichester.

Fuchs, S. 2002. Quantification of heavy metal inputs from Germany to implement the decisions of the International North Sea Protection Conference. University of Karlsruhe, Karlsruhe.

Fuhrimann, S., Stalder, M., Winkler, M. S., Niwagaba, C. B., Babu, M., Masaba, G., Kabatereine, N. B., Halage, A. A., Schneeberger, P. H., Utzinger, J. & Cissé, G. 2015. Microbial and chemical contamination of water, sediment and soil in the Nakivubo wetland area in Kampala, Uganda. Environmental Monitoring and Assessment 187: 475-489.

Fylstra, D., Lasdon, L., Watson, J. & Waren, A. 1998. Design and use of the Microsoft Excel Solver. Interfaces 28: 29-55.

Ganesh, R., Balaji, G. & Ramanujam, R. A. 2006. Biodegradation of tannery wastewater using sequencing batch reactor: respirometric assessment. Bioresource Technology 97: 1815-1821.

Gashaw, T., Bantider, A. & G/Silassie, H. 2014. Land degradation in Ethiopia: causes, impacts and rehabilitation techniques. Journal of Environmental and Earth Science 4: 98-104.

Gasparini, D., Cunha, F., Bottino, F. & Carmo, M. 2010. Land use influence on eutrophication-related water variables: case study of tropical rivers with different degrees of anthropogenic interference. Acta Limnologica Brasiliensia 22: 35-40.

Gaur, V. K., Gupta, S. K., Pandey, S. D., Gopal, K. & Misra, V. 2005. Distribution of heavy metals in sediment and water of river Gomti. Environmental Monitoring and Assessment 102: 419-433.

Gavian, S. 1999. Measuring the production efficiency of alternative land tenure contracts in a mixed crop-livestock system in Ethiopia. ILRI, Addis Ababa.

Gebeyehu, Z. H. 2013. Towards Improved Transactions of Land Use Rights in Ethiopia. In: Annual World Bank Conference on Land and Poverty 2013, Washington, D.C.

Gebrekidan, A., Gebresellasie, G. & Mulugeta, A. 2009. Environmental impacts of Sheba tannery effluents on the surrounding water bodies, Ethiopia. Bulletin of the Chemical Society of Ethiopia 23: 269 -274.

Gebreselassie, S., Kirui, O. & Mirzabaev, A. 2016. Economics of Land Degradation and Improvement in Ethiopia. In: A global assessment for sustainable development, Nkonya, E., Mirzabaev, A., von Braun, J. (Eds). Springer, Cham.

Getu, M. 2009. Ethiopian floriculture and its impact on the environment. Mizan law review 3: 240-270.

Ghaly, A. E., Ananthashankar, R., Alhattab, M. & Ramakrishnan, V. 2014. Production, characterization and treatment of textile effluents: a critical review. Journal of Chemical Engineering & Process Technology 5: 182-200.

Ghoreishi, S. M. & Haghighi, R. 2003. Chemical catalytic reaction and biological oxidation for treatment of non-biodegradable textile effluent. Chemical Engineering 95: 163-169.

Gil, K. & Kim, T. W. 2012. Determination of first flush criteria from an urban residential area and a transportation land-use area. Desalination and Water Treatment 40: 309-318.

Goher, M. E., Hassan, A. M., Abdel-Moniem, I. A., Fahmy, A. H. & El-Sayed, S. M. 2014. Evaluation of surface water quality and heavy metal indices of Ismailia Canal, Nile River, Egypt Egyptian. Journal of Aquatic Research 40: 225–233.

Gordon, M. 2005. Mapping hazard from urban non-point pollution: A screening model to support sustainable urban drainage planning. Journal of Environmental Management 74: 1-9.

Griffith, J. A. 2002. Geographic Techniques and Recent Applications of Remote Sensing to Landscape-Water Quality Studies. Water, Air & Soil Pollution 138: 181-197.

Grossman, G. & Krueger, A. 1991. Environmental impacts of a North American free trade agreement. National Bureau of Economic Research, Cambridge.

Gumbo, B. 2005. Short-cutting the phosphorus cycle in urban ecosystems. Delft University of Technology and UNESCO-IHE, Institute for Water Education, The Netherlands.

Gurung, D. P., Githinji, L. J. & Ankumah, R. O. 2013. Assessing the Nitrogen and Phosphorus Loading in the Alabama (USA) River Basin Using PLOAD Model. Air, Soil and Water Research 6: 23.

Haith, D. A., Mandel, R. & Wu, R. S. 1992. Generalized watershed loading functions version 2.0 user's manual. Cornell University, New York.

Håkanson, L. & Jansson, M. 1983. Principles of lake sedimentology. Springer, Berlin.

Halling-Sörensen, B. & Jörgensen, S. 2008. Nitrogen compounds as pollutants. Elsevier BV, Amsterdam.

Hamilton, S. 2008. Sources of Uncertainty in Canadian Low Flow Hydrometric Data. Canadian Water Resources Journal 33: 125-136.

Harmel, R., Cooper, R., Slade, R., Haney, R. & Arnold, J. 2006. Cumulative uncertainty in measured streamflow and water quality data for small watersheds. Transactions of the ASABE 49: 689-701.

Hashem, M. A., Islam, A., Mohsin, S. & Nur-A-Tomal, M. S. 2015. Green environment suffers by discharging of high-chromium-containing wastewater from the tanneries at Hazaribagh, Bangladesh. Sustainable Water Resources Management 1: 343-347.

Henze, M. & Comeau, Y. 2008. Wastewater characterization. In: Biological wastewater treatment: principles, modelling and design, Henze, M., van Loosdrecht, M., Ekama, G., Brdjanovic, D. (Eds). IWA Publishing, London.

Herschy, R. W. 1985. Streamflow measurement. CRC Press, London.

Hildebrand, L. 2002. Integrated coastal management: lessons learned and challenges ahead. Coastal Zone Canada Association, Hamilton.

Hoos, A. B. 2008. Data to support statistical modeling of instream nutrient load based on watershed attributes, Southeastern United States. U.S. Geological Survey, Reston.

Horst, M., Travel, R. & Tokarz, E. 2008. BMP pollutant removal efficiency. World Environmental and Water Resources Congress, Reston.

Hove, M., Ngwerume, E. & Muchemwa, C. 2013. The urban crisis in Sub-Saharan Africa: A threat to human security and sustainable development. Stability: International Journal of Security and Development 2: 1-14.

Ierodiaconou, D., Laurenson, L., Leblanc, M., Stagnitti, F., Duff, G., Salzman, S. & Versace, V. 2005. The consequences of land use change on nutrient exports: a regional scale assessment in south-west Victoria, Australia. Journal of Environmental Management 74: 305-316.

Ilijevic, K., Obradovic, M., Jevremovic, V. & Grzetic, I. 2015. Statistical analysis of the influence of major tributaries to the eco-chemical status of the Danube River. Environmental Monitoring and Assessment 187: 1-25.

Ilou, I., Souabi, S. & Digua, K. 2014 Quantification of pollution discharges from tannery wastewater and pollution reduction by pretreatment station. International Journal of Science and Research 3: 1706-1715.

Inyang, U. E., Bassey, E. N. & Inyang, J. D. 2012. Characterization of brewery effluent fluid. Journal of Engineering and Applied Sciences 4: 66-77.

Ipeaiyeda, A. R. & Onianwa, P. C. 2009. Impact of brewery effluent on water quality of the Olosun river in Ibadan, Nigeria. Chemistry and Ecology 25: 189-204.

Irvine, K., Allott, N., Mills, P. & Free, G. 2001. The use of empirical relationships and nutrient export coefficients for predicting phosphorus concentrations in Irish lakes. Verhandlungen des Internationalen Verein Limnologie 27: 1127-1131.

Islam, M. S., Han, S. & Masunaga, S. 2014. Assessment of trace metal contamination in water and sediment of some rivers in Bangladesh. Journal of Water and Environment Technology 12: 109-121.

ISO. 2003. Part 3: Guidance on preservation and handling of water samples (ISO 5667-3). In: Water quality sampling, Sheffer, M. (ed). International Organization for Standards, Geneva.

Jining, C. & Yi, Q. 2009. Point sources of pollution: Local effects and control. Encyclopedia of Life Support System, Tsinghua.

Johnes, P., Moss, B. & Phillips, G. 1996. The determination of total nitrogen and total phosphorus concentrations in freshwaters from land use, stock headage and population data: testing of a model for use in conservation and water quality management. Freshwater Biology 36: 451-473.

Johnes, P. J. 1996. Evaluation and management of the impact of land use change on the nitrogen and phosphorus load delivered to surface waters: the export coefficient modelling approach. Journal of Hydrology 183: 323-349.

Johnson, L., Richards, C., Host, G. & Arthur, J. 1997. Landscape influences on water chemistry in midwestern stream ecosystems. Freshwater Biology 37: 193-208.

Jönsson, H., Baky, A., Jeppsson, U., Hellström, D. & Kärrman, E. 2005. Composition of urine, feaces, greywater and biowaste for utilisation in the URWARE model. Urban Water, Chalmers University of Technology, Göteborg.

Jumbe, A. & Nandini, N. 2009. Heavy metals analysis and sediment quality values in urban lakes. American Journal of Environmental Sciences 5: 678-687.

JWQB (Japan Water Quality Bureau) 1998. Water environment management in Japan. Japan Water Quality Bureau, Tokyo.

Kamiya, H., Kano, Y., Mishima, K., Yoshioka, K., Mitamura, O. & Ishitobi, Y. 2008. Estimation of long-term variation in nutrient loads from the Hii River by comparing the change in observed and calculated loads in the catchments. Landscape and Ecological Engineering 4: 39-46.

Karki, R., Tagert, M. L. M., Paz, J. O. & Bingner, R. L. 2017. Application of AnnAGNPS to model an agricultural watershed in East-Central Mississippi for the evaluation of an on-farm water storage (OFWS) system. Agricultural Water Management 192: 103-114.

Karrari, P., Mehrpour, O. & Abdolahi, M. 2012. A systematic review on status of lead pollution and toxicity in Iran; guidance for preventive measures. DARU Journal of Pharmaceutical Sciences 20: 2-19.

Katiyar, S. 2011. Impact of tannery effluent with special reference to seasonal variation on physico-chemical characteristics of river water at Kanpur, India. Journal of Environmental & Analytical Toxicology 1: 35-42.

Kato, T., Kuroda, H. & Nakasone, H. 2009. Runoff characteristics of nutrients from an agricultural watershed with intensive livestock production. Journal of Hydrology 368: 79-87

Katsriku, F., Wilson, M., Yamoah, G., Abdulai, J., Rahman, B. & Grattan, K. 2015. Framework for time relevant water monitoring system. In: Computing in Research and Development in Africa, Gamatié, A. (ed). Springer International Publishing Cham.

Kebede, B. 2002. Land tenure and common pool resources in rural Ethiopia: a study based on fifteen sites. African Development Review 14: 113-149.

Kelderman, P. 2012. Sediment pollution, transport, and abatement measures in the city canals of Delft, The Netherlands. Water, Air and Soil Pollution 223: 4627-4645.

Kelderman, P., Koech, D., Gumbo, B. & O'Keeffe, J. 2009. Phosphorus budget in a low-income, peri-urban area of Kibera in Nairobi (Kenya). Water Science and Technology 60 2669-2676.

Kennedy, A. E. 1984. Discharge ratings at gaging stations. US Government Printing Office, Washington, D.C.

Kihampa, C. 2013. Heavy metal contamination in water and sediment downstream of municipal wastewater treatment plants, Dar es Salaam, Tanzania. International Journal of Environmental Sciences 3: 1407-1415.

Kimura, S., Liang, L. & Hatano, R. 2004. Influence of long-term changes in nitrogen flows on the environment: a case study of a city in Hokkaido, Japan. Nutrient Cycling in Agroecosystems 70: 271-282.

Kishe, M. & Machiwa, J. 2003. Distribution of heavy metals in sediments of Mwanza gulf of lake Victoria, Tanzania. Environment International 28: 619-625.

Kombolcha Meteorological Branch Directorate. 2015. Meteorological information and climate records of Kombolcha adiminstration city. KMBD, Kombolcha.

Kumpel, E., Peletz, R., Bonham, M., Fay, A., Cock-Esteb, A. & Khush, R. 2015. When are mobile phones useful for water quality data collection? An analysis of data flows and ict applications among regulated monitoring institutions in sub-Saharan Africa. International Journal of Environment Research and Public Health 12: 10846–10860.

Landgrebe, D. 1998. Multispectral data analysis: a signal theory perspective. Purdue University, West Lafayette.

Landner, L. & Reuther, R. 2004. Metals in society and in the environment: A critical review of current knowledge on fluxes, speciation, bioavailability and risk for adverse effects of copper, chromium, nickel and zinc. Springer, Dordrecht.

Larissa, D., Mesfin, M., Elias, D., Carlos, E. & Veiga, C. 2013. Assessment of heavy metals in water samples and tissues of edible species from Awassa and Koka rift valley lakes, Ethiopia. Environmental Monitoring and Assessment 185: 3117-3131.

Lehner, B., Verdin, K. & Jarvis, A. 2008. New global hydrography derived from spaceborne elevation data. Eos, Transactions of the American Geophysical Union 89: 93-94.

Lin, J. & Chen, S. 1998. The relationship between adsorption of heavy metal and organic matter in river sediments. Environment International 24: 345-352.

Lin, J. P. 2004. Review of Published Export Coefficient and Event Mean Concentration (EMC) Data. U.S. Army Engineer Research and Development Center, Vicksburg.

Lin, J. P. & Kleiss, B. A. 2004. Availability of a PowerPoint-based tutorial on applying PLOAD for wetlands management. U.S. Army Engineer Research and Development Center, Vicksburg.

Liu, R., Yang, Z., Shen, Z., Yu, S., Ding, X., Wu, X. & Liu, F. 2009. Estimating nonpoint source pollution in the upper Yangtze river using the export coefficient model, remote sensing, and geographical information system. Journal of Hydraulic Engineering 135: 698-704.

Loehr, C., Ryding, O. & Sonzogni, C. 1989. Estimating the nutrient load to a waterbody. In: The control of eutrophication of lakes and reservoirs, Ryding, S., Rast, W. (Eds), Paris.

Loucks, P., Beek, E., Stedinger, J., Dijkman, J. & Villars, M. 2005. Water resources systems planning and management : An introduction to methods, models and applications. UNESCO, Paris.

Macdonald, D., Berger, T., Wood, K., Brown, J., Johnsen, T., Haines, M., Brydges, K., MacDonald, M., Smith, S. & Shaw, D. 2000a. A compendium of environmental quality benchmarks. MacDonald Environmental Sciences Limited, Vancouver.

MacDonald, D., Ingersoll, C. & Berger, T. 2000b. Development and evaluation of consensus-based sediment quality guidelines for freshwater ecosystems. Archives Of Environmental Contamination and Toxicology 39: 20-31.

Majumdar, J., Baruah, B.K. and Dutta, K. . 2007. Sources and characteristics of galvanizing industry effluent. Journal of Industrial Pollution Control 23 119-123.

Manzoor, S., Shah, M. H., Shaheen, N., Khalique, A. & Jaffar, M. 2006. Multivariate analysis of trace metals in textile effluents in relation to soil and groundwater. Journal of Hazardous Materials 137: 31-37.

McDowell, R. W. 2008. Environmental impact of pasture-based farming. CABI, Wallingford.

McFarland, A. & Hauck, L. 2001. Determining nutrient export coefficients and source loading uncertainty using in-stream monitoring data. Journal of American Water Resources Association 37: 223-236.

Mesfin, M. 2012. Industrial zones development corporation wins formation approval. Berhanena Selam Printing Press, Addis Ababa.

134

Meynendonckx, J., Heuvelmans, G., Muys, B. & Feyen, J. 2006. Effects of watershed and riparian zone characteristics on nutrient concentrations in the River Scheldt Basin. Hydrology and Earth System Sciences Discussions 3: 653-679.

MoFED. 2002. Ethiopia: sustainable development and poverty reduction Program MinIstry of Finance and Economic Development, Addis Ababa.

Moges, A. M., Tilahun, A. T., Ayana, E. K., Moges, M. M., Gabye, N., Giri, S. & Steenhuis, T. S. 2016. Non-point source pollution of dissolved phosphorus in the Ethiopian highlands: The Awramba watershed near Lake Tana. CLEAN Soil Air Water 44: 703-709

Mohammed, S. 2003. A review of water quality and pollution studies in Tanzania. Journal of the Human Environment 31: 617-620.

Mourad, D. S. J. 2008. Patterns of nutrient transfer in lowland catchments : a case study from northeastern Europe. Koninklijk Nederlands Aardrijkskundig Genootschap, Utrecht.

Mustapha, A. & Aris, A. Z. 2012. Spatial aspects of surface water quality in the Jakara Basin, Nigeria using chemometric analysis. Environmental Science and Health 47: 1455–1465.

Mwinyihija, M., Meharg, A., Dawson, J., Strachan, N. J. C. & Killham, K. 2006. An Ecotoxicological Approach to Assessing the Impact of Tanning Industry Effluent on River Health. Archives of environmental contamination and toxicology 50: 316-324.

Nagpal, N., Pommen, L. & Swain, L. 1995. Approved and working criteria for water quality. Ministry of Environment, Victoria.

NASA. 2014. Landsat 8: LC81680522014290-SC20150309084744, Level1, Terrain Corrected. In: Landsat Program USGS, Sioux Falls, USA.

Ndimele, P. E., Pedro, M. O., Agboola, J. I., Chukwuka, K. S. & Ekwu, A. O. 2017. Heavy metal accumulation in organs of Oreochromis niloticus (Linnaeus, 1758) from industrial effluent-polluted aquatic ecosystem in Lagos, Nigeria. Environmental Monitoring and Assessment 189: 255-267.

Negassa, A. & Jabbar, M. 2008. Livestock ownership, commercial off-take rates and their determinants in Ethiopia. ILRI, Nairobi.

Nile Basin Initiative. 2016. Basin Monitoring. In: The Nile basin: water resources atlas, Jan, H., Ahmed, K., Emmanuel, O. (Eds). New Vision Printing and Publishing Company Ltd, Kampala.

Niño de Guzmán, G. T., Hapeman, C. J., Prabhakara, K., Codling, E. E., Shelton, D. R., Rice, C. P., Hively, W. D., McCarty, G. W., Lang, M. W. & Torrents, A. 2012. Potential pollutant sources in a Choptank River (USA) subwatershed and the influence of land use and watershed characteristics. Science of the Total Environment 430: 270-279.

Noto, L. V., Ivanov, V. Y., Bras, R. L. & Vivoni, E. R. 2008. Effects of initialization on response of a fully-distributed hydrologic model. Journal of Hydrology 352: 107-125.

Novotny, V. & Chesters, G. 1981. Handbook of nonpoint pollution: sources and management. Van Nostrand Reinhold, New York.

Nriagu, J. O. & Pacyna, J. M. 1988. Quantitative assessment of worldwide contamination of air, water and soils by trace metals. Nature 333: 134-139.

Nyamangara, J., Bangira, C., Taruvinga, T., Masona, C., Nyemba, A. & Ndlovu, D. 2008. Effects of sewage and industrial effluent on the concentration of Zn, Cu, Pb and Cd in water and sediments along Waterfalls stream and lower Mukuvisi River in Harare, Zimbabwe. Physics and Chemistry of the Earth 33: 708-713.

Nyenje, P. M., Foppen, J. W., Uhlenbrook, S., Kulabako, R. & Muwanga, A. 2010. Eutrophication and nutrient release in urban areas of sub-Saharan Africa: a review. Science of the Total Environment 408: 447-455.

OECD (Organisation for Economic Co-operation and Development) 1999. Environmental requirements for industrial permitting. OECD Publishing, Paris.

OECD. 2007. Environment and regional trade agreements. OECD Publishing, Paris.

OECD. 2013. Compendium of Agri-environmental Indicators. OECD Publishing, Paris.

Oguttu, H. W., Bugenyi, F. W., Leuenberger, H., Wolf, M. & Bachofen, R. 2008. Pollution menacing Lake Victoria: quantification of point sources around Jinja Town, Uganda. Water SA 34: 89-98.

Ohioma, I., Obejesi, N. L. & Amraibure, O. 2009. Studies on the pollution potential of wastewater from textile processing factories in Kaduna, Nigeria Journal of Toxicology and Environmental Health Sciences 1: 034-037.

Ometo, J., B., P. H., Martinelli, L., Ballester, M. V., Gessner, A. L., Krusche, A., Victoria, R. L. & Williams, M. 2000. Effects of land use on water chemistry and macroinvertebrates in two streams of the Piracicaba river basin, South-East Brazil. Freshwater Biology 44: 327-337.

Ongley, E. D. 1993. Global water pollution: challenges and opportunities. SIWI, Stockholm.

Ongley, E. D. & Booty, W. G. 1999. Pollution Remediation Planning In Developing Countries. Water International 24: 31-38.

Ottens, J. J., Claessen, F. A., Stoks, P. G., Timmerman, J. G. & Ward, R. C. 1997. Monitoring and assessment in water management. Institute for Inland Water Management and Waste Water Treatment, Nunspeet.

Oyewo, E. & Don-Pedro, K. 2009. Estimated annual discharge rates of heavy metals from industrial sources around Lagos; a West African Coastal Metropolis. West African Journal of Applied Ecology 4: 115-123.

Packett, R., Dougall, C., Rohde, K. & Noble, R. 2009. Agricultural lands are hot-spots for annual runoff polluting the southern Great Barrier Reef lagoon. Marine Pollution Bulletin 58: 976-986.

Pacyna, J. M. & Pacyna, E. G. 2001. An assessment of global and regional emissions of trace metals to the atmosphere from anthropogenic sources worldwide. Environmental Reviews 9: 269-298.

Pagano, M. & Gauvreau, K. 2000. Principles of biostatistics. Duxbury, Pacific Grove.

Parawiraa, W., Kudita, I., Nyandoroh, M. G. & Zvauya, R. 2005. A study of industrial anaerobic treatment of opaque beer brewery wastewater in a tropical climate using a full-scale UASB reactor seeded with activated sludge. Process Biochemistry 40: 593-599.

Pawlikowski, M., Szalinska, E., Wardas, M. & Dominik, J. 2006. chromium originating from tanneries in river sediments: a preliminary investigation from the upper dunajec river (Poland). Polish Journal of Environmental Studies 15: 885-894.

Peletz, R., Kisiangani, J., Bonhama, M., Ronoh, P., Delaire, C., Kumpela, E., Marks, S. & Khush, R. 2018. Why do water quality monitoring programs succeed or fail? A qualitative comparative analysis of regulated testing systems in sub-Saharan Africa. International. Journal of Hygiene and Environmental Health 221: 907-920.

Peletz, R., Kumpel, E., Bonham, M., Rahman , Z. & Khush, R. 2016. To what extent is drinking water tested in Sub-Saharan Africa? A comparative analysis of regulated water quality monitoring. International Journal of Environment Research and Public Health 13: 275-288.

Piguet, F. 2002. Cheffa valley: refuge for 50000 pastorialist and 200000 animals: report on present humanterian situation and livestock conditions in selected areas in and arround Afar region. UN Emergency Unit for Ethiopia Addis Ababa.

Pourkhabbaz, A., Khazaei, T., Behravesh, S., Ebrahimpour, M. & Pourkhabbaz, H. 2011. Effect of Water Hardness on the Toxicity of Cobalt and Nickel to a Freshwater Fish, Capoeta fusca. Biomedical and Environmental Sciences 24: 656-660.

Prabu, P. C. 2009. Impact of heavy metal contamination of Akaki river of Ethiopia on soil and metal toxicity on cultivated vegetable crops. Electronic Journal of Environmental Agricultural and Food Chemistry 8: 818 - 827.

Preston, S. D., Bierman, V. J., Silliman, S. E., Geological, S. & Purdue University. Water Resources Research, C. 1989. Evaluation of methods for the estimation of tributary mass loading rates. Water Resources Research Center, Purdue University, West Lafayette.

Quilbé, R., Rousseau, A. N., Duchemin, M., Poulin, A., Gangbazo, G. & Villeneuve, J.-P. 2006. Selecting a calculation method to estimate sediment and nutrient loads in streams: application to the Beaurivage River (Québec, Canada). Journal of Hydrology 326: 295-310.

R Core Team. 2015. R: A language and environment for statistical computing. R Foundation for Statistical Computing, Vienna.

Rajaram, T. & Das, A. 2008. Water pollution by industrial effluents in India: Discharge scenarios and case for participatory ecosystem specific local regulation. Futures 40: 56-69.

Rast, W. & Lee, G. 1983. Nutrient Loading Estimates for Lakes. Journal of Environmental Engineering 109: 502-517.

Reggiani, P. & Schellekens, J. 2003. Modelling of hydrological responses: the representative elementary watershed approach as an alternative blueprint for watershed modelling. Hydrological Processes 17: 3785-3789.

Rice, E. W., Baird, R. B., Eaton, A. D. & Clesceri, L. S. 2012. Standard methods for examination of water and wastewater. American Public Health Association, Washington D.C.

RIDEM. 1997. Water Quality Regulation. Rhode Island Department of Environmental Management, Division of Water Resources, Rhode Island.

Ritchie, C. J., Zimba, P. & Everitt, H. J. 2003. Remote Sensing Techniques to Assess Water Quality. Photogrammetric Engineering and Remote Sensing 69: 695-704.

138

Rode, M., Kebede, T., Arhonditsis, G., Balin, D., Krysanova, V., Van Griensven, A. & Van Der Zee, S. 2010. New challenges in integrated water quality modelling. Hydrological Processes 24: 3447-3461.

Rose, C., Parker, A., Jefferson, B. & Cartmell, E. 2015. The Characterization of Feces and Urine: A Review of the Literature to Inform Advanced Treatment Technology. Critical Reviews in Environmental Science and Technology 45: 1827-1879.

Rudi, L. M., Azadi, H. & Witlox, F. 2012. Reconcilability of socio-economic development and environmental conservation in Sub-Saharan Africa. Global and Planetary Change 86-87: 1-10.

Ruffeis, D., Loiskandl, W., Awulachew, S. & Boelee, E. 2010. Evaluation of the environmental policy and impact assessment process in Ethiopia. Impact Assessement and Project Appraisal 28: 29-40.

Rungnapa, T., Athiwatr, J., Chantana, Y. & Thumrongrut, M. 2010. Analysis of steel production in Thailand: Environmental impacts and solutions. Energy 35: 4192-4200.

Sabo, A., Gani, A. M. & Ibrahim, A. 2013. Pollution status of heavy metals in water and bottom sediment of River Delimi in Jos Nigeria. American Journal of Environmental Protection 1: 47-53.

Salomons, W. & Förstner, U. 1984. Metals in the hydrocycle. Springer-Verlag, Berlin.

Satyawali, Y. & Balakrishnan, M. 2008. Wastewater treatment in molasses-based alcohol distilleries for COD and color removal: A review. Journal of Environmental Management 86: 481-497.

Scheren, P., Zanting, H. & Lemmens, A. 2000. Estimation of water pollution sources in Lake Victoria, East Africa: application and elaboration of the rapid assessment methodology. Journal of Environmental Management 58: 235-248.

Schnurbusch, S. A. 2000. A mixing zone guidance document prepared for the Oregon department of environmental quality. Portland State University, Portland.

Schwarz, G., Hoos, A., Alexander, R. & Smith, R. 2006. The SPARROW surface water-quality model: theory, application and user documentation. US geological survey techniques and methods report, Washington D.C.

Shapiro, S. S. & Wilk, M. B. 1965. An analysis of variance test for normality (complete samples). Biometrika 52: 591- 611.

Shaver, E., Horner, R., Skupien, J., May, C. & Ridley, G. 2007. Fundamentals of Urban Runoff Management: Technical and Institutional Issues. North American Lake Management Society, Madison.

Shrestha, S., Kazama, F., Newham, L. T. H., Babel, M. S., Clemente, R. S., Ishidaira, H., Nishida, K. & Sakamoto, Y. 2008. Catchment scale modelling of point source and non-point source pollution loads using pollutant export coefficients determined from long-term in-stream monitoring data. Journal of Hydro-environment Research 2: 134-147.

Sial, R. A., Chaudhary, M. F., Abbas, S. T., Latif, M. I. & Khan, A. G. 2006. Quality of effluents from Hattar industrial estate. Journal of Zhejiang University Science 7: 974-980.

Sikder, M. T., Kihara, Y., Yasuda, M., Mihara, Y., Tanaka, S., Odgerel, D., Mijiddorj, B., Syawal, S. M., Hosokawa, T. & Saito, T. 2013. River water pollution in developed and developing countries: Judge and assessment of physicochemical characteristics and selected dissolved metal concentration. CLEAN Soil Air Water 41: 60-68.

Singh, V. P. 1995. Watershed modeling. Water Resources Publications, Highlands Ranch.

Sliva, L. & Dudley, W. D. 2001. Buffer Zone versus Whole Catchment Approaches to Studying Land Use Impact on River Water Quality. Water Research 35: 3462-3472.

Smith, L. E. D. & Siciliano, G. 2015. A comprehensive review of constraints to improved management of fertilizers in China and mitigation of diffuse water pollution from agriculture. Agriculture, Ecosystems and Environment 209: 15-25.

Soranno, P. A., Cheruvelil, K. S., Wagner, T., Webster, K. E. & Bremigan, M. T. 2015. Effects of land use on lake nutrients: The importance of scale, hydrologic connectivity, and region. PloS one 10.

Soranno, P. A., Hubler, S. L., Carpenter, S. R. & Lathrop, R. C. 1996. Phosphorus loads to surface waters: A simple model to account for spatial pattern of land use. Ecological Applications 6: 865-878.

Steel, W. F. & Evans, J. W. 1984. Industrialization in sub-Saharan Africa: strategies and performance World Bank, Washington D.C.

Stigliani, W. M., Jaffe, P. R. & Anderberg, S. 1993. Heavy metal pollution in the Rhine Basin. Environmental Science and Technology 27: 786-793.

Su, S., Xiao, R., Mi, X., Xu, X., Zhang, Z. & Wu, J. 2013. Spatial determinants of hazardous chemicals in surface water of Qiantang River, China. Ecological Indicators 24: 375–381.

Swartz, R. C. 1999. Consensus sediment quality guidelines for PAH mixtures. Environmental Toxicololgy and Chemistry 18: 780 –787.

Taddese, G. 2001. Land degradation: a challenge to Ethiopia. Environmental management 27: 815-824.

Taebi, A. & Droste, R. L. 2004. First flush pollution load of urban stormwater runoff Journal of Environmental Engineering and Science 3: 301-309.

Tariq, S. R., Shah, M. H., Shaheen, N., Khalique, A., Manzoor, S. & Jaffar, M. 2006. Multivariate analysis of trace metal levels in tannery effluents in relation to soil and water: a case study from Peshawar, Pakistan. Journal of Environmental Management 79: 20 -29.

Tasdighi, A., Arabi, M. & Osmond, D. L. 2017. The relationship between land use and vulnerability to nitrogen and phosphorus pollution in an urban watershed. Journal of Environmental Quality 46: 113-122.

Tefera, B. & Sterk, G. 2010. Land management, erosion problems and soil and water conservation in Fincha's watershed, Western Ethiopia. Land Use Policy 27: 1027-1037.

Tucker, G. E. & Bras, R. L. 1998. Hillslope processes, drainage density, and landscape morphology. Water Resources Research 34: 2751-2764.

UNDP. 2015. World leaders adopt sustainable development goals. United Nations Development Programme, New York.

UNEP. 2004. Global environment outlook scenario framework: Background paper for UNEP'S third global environment outlook report (GEO-3). UNEP, Nairobi.

United Nations. 2016. Paris Agreement United Nations, NewYork.

USEPA. 1986. Quality criteria for water. EPA, Washington D.C.

USEPA. 1997a. The incidence and severity of sediment contamination in surface waters of the United States. In: National Sediment Quality Survey. USEPA, Washington D.C.

USEPA. 1997b. Recent developments for in-situ treatment of metals contaminated soils. U.S. Environmental Protection Agency, Washington D.C.

USEPA. 1998. National recommended water quality criteria: Republication. Office of Water, United States Environmental Protection Agency Washington D.C.

USEPA. 2008. Handbook for developing watershed plans to restore and protect our waters. U.S. Environmental Protection Agency, Washington D.C.

USEPA. 2014. Industrial effluent guidelines. US Environmental Protection Agency, Washington D.C.

USEPA. 2015. BASINS 4.1 (better assessment science integrating point and non-point sources): modeling framework. National Exposure Research Laboratory, Quezon.

USGS. 2006. Shuttle Radar Topography Mission, 1 Arc-second scene SRTM1N11E039V3, void filled. Global Land Cover Facility, University of Maryland, College Park.

Van der Perk, M. 2006. Soil and water contamination: from molecular to catchment scale. Taylor and Francis-Balkema, Leiden.

Velthof, G. L., Lesschen, J. P., Webb, J., Pietrzak, S., Miatkowski, Z., Pinto, M. & Oenema, O. 2014. The impact of the nitrates directive on nitrogen emissions from agriculture in the EU-27 during 2000–2008. Science of the Total Environment 468: 1225-1233.

Vink, R. & Behrendt, H. 2002. Heavy metal transport in large river systems: heavy metal emissions and loads in the Rhine and Elbe river basins. Hydrological Processes 16: 3227-3244.

Voien, S. 1998. Environmental management with ISO 14000. Sage Publications, Geneva.

Walker, W. W. 1987. Empirical Methods for Predicting Eutrophication in Impoundments. Report 4. Phase III. Applications Manual. Johnson Publishing Co, Chicago.

Walker, W. W. 1990. FLUX stream load computations. Version 4.4. United States Army Corps of Engineering Waterways Exp, Vicksburg.

Walker, W. W. 1999. Simplified procedures for eutrophication assessment and prediction user manual. United States Army Corps of Engineers, Vicksburg.

Wang, L., Wang, W. D., Gong, Z. G., Liu, Y. L. & Zhang, J. J. 2006. Integrated management of water and ecology in the urban area of Laoshan district, Qingdao, China. Ecological Engineering 27: 79-83.

Wang, Q., Li, S., Jia, P., Qi, C. & Ding, F. 2013. A review of surface water quality models. The Scientific World Journal 2013: 1-7.

Warn, T. 2010. SIMCAT 11.5 A Guide and Reference for Users. Environment Agency, London.

Wenger, S. J. 1999. A review of the scientific literature on riparian buffer width, extent and vegetation. Institute of Ecology, Athens.

Whitehead, C. 1988. European Community environmental legislation 1967-1987. Commission of the European Communities, Brussels.

Whitehead, P. G., Williams, R. J. & Lewis, D. R. 1997. Quality simulation along river systems (QUASAR): Model theory and development. Science of the Total Environment 194-195: 447.

WHO. 2011. Guidelines for drinking-water quality. World Health Organization, Geneva.

WHO/UNICEF. 2014. Progress on drinking water and sanitation: 2014 update. WHO, Geneva.

Wood, M. S. & Beckwith, M. A. 2008. Coeur d'Alene Lake, Idaho: Insights gained from limnological studies of 1991–92 and 2004–06 USGS, Reston.

World Bank 2015. Enhancing shared prosperity through equitable services: environmental and social systems assessment. World Bank Group, Washington D.C.

Xu, X., Zhao, Y., Zhao, X., Wang, Y. & Deng, W. 2014. Sources of heavy metal pollution in agricultural soils of a rapidly industrializing area in the Yangtze Delta of China. Ecotoxicology and Environmental Safety 108: 161-167.

Yabe, J., Ishizuka, M. & Umemura, T. 2010. Current levels of heavy metal pollution in Africa. Journal of Veterinary Medical Science 72: 1257-1263.

Yetunde, J. 2006. Export coefficients for total phosphorus, total nitrogen and total suspended solids in the southern Alberta region, a review of literature. Information Centre Alberta Environment, Edmonton.

Yi, Y., Yang, Z. & Zhang, S. 2011. Ecological risk assessment of heavy metals in sediment and human health risk assessment of heavy metals in fishes in the middle and lower reaches of the Yangtze River basin. Environmental Pollution 159: 2575–2585.

Young, D. J. 2010. Development of an ArcGIS-pollutant load application (PLOAD) tool. Texas A&M University, Texas.

Yuan, G., Liu, C., Chen, L. & Yang, Z. 2011. Inputting history of heavy metals into the inland lake recorded in sediment profiles: Poyang lake in China. Journal of hazardous materials 185: 336–345.

Yusuff, R. O. & Sonibare, J. A. 2004. Characterization of textile industries effluents in Kaduna, Nigeria and pollution implications. Global Nest Journal 6: 212-221.

Zhenyao, S., Qian, H., Zheng, C. & Yongwei, G. 2011. A framework for priority non-point source area identification and load estimation integrated with APPI and PLOAD model in Fujiang Watershed, China. Agricultural Water Management 98: 977–989.

Zinabu, E. 2011. Assessment of the impact of industrial effluents on the quality of irrigation water and changes in soil characteristics: the case of Kombolcha town. Irrigation and Drainage 60: 644-653.

Zinabu, E., Kelderman, P., van der Kwast, J. & Irvine, K. 2017a. Impacts and policy implications of metals effluent discharges into rivers within industrial zone: a sub-Saharan perspective from Ethiopia. Environmental Management 61: 700-715.

Zinabu, E., Kelderman, P., van der Kwast, J. & Irvine, K. 2018. Evaluating diffuse and point source nutrient transfers in Kombolcha River Basin, an industrializing Ethiopian catchment. Land Degradation and Development 29: 3366-3378.

Zinabu, E., van der Kwast, J., Kelderman, P. & Irvine, K. 2017b. Estimating total nitrogen and phosphorus losses in a data-poor Ethiopian catchment. Journal of Environmental Quality 46: 1519-1525.

Samenvatting

In de hydrologie van stroomgebieden van rivieren, is het bepalen van de relatieve belastngen van diffuse- en puntbronnen van zware metalen en nutriënten, alsmede van de verschillende hydrologische stromingen, een belangrijke onderzoeksuitdaging. Inzicht in de overdracht, belastingen en concentraties van deze belastingen in stroomgebieden is nuttig voor het opzetten en implementeren van beleid op het gebied van oppervlaktewaterbeheer. In landen ten zuiden van de Sahara zijn slechts enkele studies uitgevoerd op het gebied van bovenstaande aspecten. Zelfs voor kleine stroomgebieden is het in het algemeen moeilijk om hydrologische en hydro-chemische gegevens te verkrijgen. Dit proefschrift richt zich op het bepalen van zware metalen- en nutriëntenbelastingen van industrieën en landgebruik, in twee rivieren, die door een gebied in industriële ontwikkeling, Kombolcha, stromen. Ook wordt de selectie en toepassing van een model voor totaal-stikstof en –fosfor in het Kombolcha stroomgebied besproken. De studie naar de overdracht van vervuilende stoffen afkomstig van diffuse- en puntbronnen, leidt tot ophelderen van onderzoeksvragen op het gebied van beheer, benodigde middelen en beleid, voor het monitoren van de waterkwaliteit in Ethiopische rivieren. Tevens is de studie relevant voor andere landen ten zuiden van de Sahara.

De studie werd uitgevoerd in het semi-aride stroomgebied van de stad Kombolcha, in een urbane en peri-urbane omgeving in noord-centraal Ethiopië. De rivieren Leyole en Worka lopen door de stroomgebieden, en ontvangen industriële effluenten van verschillende fabrieken, alsmede afspoeling van het omliggende stroomgebied. De rivieren stromen naar de grotere Borkena rivier. Het doel van dit onderzoek was het monitoren en kwantificeren van bronnen en overdracht van zware metalen (Cr, Cu, Zn en Pb) en nutriënten ((NH$_4$ + NH$_3$ –N), NO$_3$ –N, TN, PO$_4$ –P, TP), in de Leyole en Worka, alsmede het evalueren van beheersaspecten, in een stroomgebied met beperkt beschikbare gegevens. Ook werden de relatieve bijdragen van totaal-N en -P-belastingen berekend, van diffuse- en puntbronnen. De studie is geplaatst in een beheerscontext, door een overzichtsstudie van relevant beheer in Ethiopië, en in het bredere perspectief of Afrika ten zuiden van de Sahara.

De eerste meetreeks werd uitgevoerd op effluenten van vijf industrieën met, in totaal, 40 effluentmonsters in zowel 2013 als 2014. In de tweede meetreeks werd gekeken naar oppervlaktewater en sedimenten. Hierin werden 120 watermonsters verzameld van het rivierwater, in het natte seizoen van de meetjaren 2013 en 2014. Er werden in totaal 18 sedimentmonsters van de rivierbodem verzameld, op zes stations en drie data in het natte seizoen van bovengenoemde twee jaren. Ook werden in de meetcampagne in 2013 en 2014 de dagelijkse stroomdiepten in de twee rivieren, twee keer per dag, opgemeten. Hiermee konden de verdunningscapaciteiten van het rivierwater worden bepaald, aan de hand van het grafisch plotten van rivierwaterafvoer (m^3/sec.) vs. water diepte.

Met mediane concentraties van Cr in het effluent van de leerlooierij, en van Zn in de metaalverwerking, respectievelijk 26,6 en 155,750 µg/L, werden emmisierichtlijnen sterk overschreden. In de Leyole zelf werden hoge Cr waarden in het water (mediaan: 60 µg/L) en sediment (maximum: 740 mg/kg) gevonden. Cu in het rivierwater was het hoogste midstrooms in de Leyole (mediaan: 63 µg/L), maar een maximum sedimentgehalte van 417 mg/kg werd stroomopwaarts gevonden. Zn concentraties waren het hoogst bovenstrooms in het Leyole water (mediaan 521 µg/L) en sediment (maximum 36,600 mg/kg). In beide rivieren werden lage Pb gehalten gevonden, met relatief hogere waarden (maximum 3640 mg/kg) bovenstrooms in de Leyole sedimenten. Cr liet dergelijke trends zien, met verhoogde waarden in het benedenstroomse gedeelte van de Leyole. Met uitzondering van Pb overschreden alle zware metaalgehalten de richtlijnen voor aquatisch leven, drinkwaterkwaliteit, irrigatie en water voor vee. Alle zware metalen overschreden de richtlijnen voor de sedimentkwaliteit voor aquatische organismen.

Wat betreft nutriënten, de emissies van een brouwerij en vleesverwerkend bedrijf hadden hoge gehalten, met mediane concentraties TN = 21,000 – 44,000 µg/L en TP = 20,000 – 58,000 µg/L, respectievelijk gemiddeld 10 en 13 % van de totale nutriëntenbelasting. In het water werden hogere TN concentraties gevonden afkomstig van van sub-stroomgebieden met als hoogste landgebruik: landbouw, terwijl hoogste TP waarden verbonden waren met sub-stroomgebieden met heuvelachtige landschappen en boslanden. Zowel de TN en TP concentraties overschreden de richtlijnen voor bescherming van aquatische leven, irrigatie, en water voor vee.

Specifieke criteria voor de geschiktheid van een model leidden tot the gebruik van PLOAD. Dit model vertrouwt op schattingen van nutriëntenbelastingen van puntbronnen, zoals industrieën, en op exportcoëfficiënten voor landgebruik, de laatste gecalibreerd met gemeten TN en TP belastingen van de stroomgebieden. Het model werd gecalibreerd en de presentatie verhoogd, waarbij de som van de fouten met respectievelijk 89% (TN) en 5% (TP) kon worden gereduceerd. De resultaten werden gevalideerd met onafhankelijke veldgegevens.

De resultaten van dit onderzoek laten hoge belastingen zien van zware metalen en nutriënten in de rivieren van Kombolcha, een gebied in industriële ontwikkeling. Hierbij werden ook hiaten geïdentificeerd bij het bepalen van de vervuiling door zware metalen en nutriënten, en bij het implementeren van beleid. Voor toekomstig onderzoek en beheersontwikkeling wordt aanbevolen een aantal essentiële hiaten aan te pakken wat betreft punt- en diffuse belastingen van zware metalen en nutriënten uit verschillende bronnen, verbetering van landbeheer, en monitoren en regulatie.

148

About the author

Mr Eskinder Zinabu was born on June 20, 1974, in DebreBrihan, Ethiopia, and grew up in the different towns across the country. He attended his primary school in St. Ruphael School at Gullele district, Addis Ababa, and completed his secondary school at Wonji, Oromia state in Ethiopia. From 1991-1996, Mr Eskinder studied his Bachelor of Science in "*Agricultural Engineering*" at Haramaya University and acquired knowledge and skills on engineering principles in agricultural activities, and develop basic know how in Planning and designing of irrigation, soil and water conservation structures, management and selection of agricultural machineries. He was then employed at different organizations and worked at different capacity. Next, he attended Master of Science education, titled "*Tropical Land Resources Management*" in Mekelle University, from 2005 to 2008 and gained knowledge and skill on integrated environmental protection, catchment managements, and land evaluations. His Thesis work reflects on the effects of industrial effluents on irrigation water quality and changes in soil characteristics in the industrial city of Kombolcha, Ethiopia.

From 2002 to 2009, Mr Eskinder worked at the Kombolcha Agricultural College and Samara University. Currently, starting form 2009, he has been working in Wollo University as both lecturer and researcher. He is involving in personal research and collaborative developmental projects and consultation activities related to water and environmental topics, and responsible in establishing collaborative links outside the university with industrial, commercial and public organizations. Mr. Eskinder PhD programme was a sandwich framework whereby he conducted his field work in Ethiopia in conjunction with Wollo University, whereas the course works, consultation, and some part of writing of the Thesis took place at IHE-Delft. During his PhD tenure he also guided two MSc students during proposal development, field works in the Kombolcha (Ethiopia) and Thesis writing. Mr. Eskinder is a member of various professional organisations like the *International Solid Wastes Associations (ISWA), International Associations of Hydrological Sciences (IAHS), and Ethiopian Environmental Protection Society (EEPS)*.

Reputable international journals

1. **Zinabu Eskinder**, Kelderman Peter, Van der Kwast Johannes, Irvine Kenneth (2018) Evaluating the effect of diffuse and point source nutrient transfers on water quality in

150

the Kombolcha River Basin, an industrializing Ethiopian catchment. *Land Degradation and Development* 29:3366-3378. doi: https://doi.org/10.1002/ldr.3096

2. **Zinabu Eskinder**, Kelderman P, van der Kwast J, Irvine K (2018) Impacts and Policy Implications of Metals Effluent Discharges into Rivers within Industrial Zone: A Sub-Saharan Perspective from Ethiopia. *Environmental Management* 61:700-715. doi: https://doi.org/10.1007/s00267-017-0970-9

1. **Zinabu Eskinder**, van der Kwast J, Kelderman P, Irvine K (2017) Estimating total nitrogen and phosphorus losses in a data-poor Ethiopian catchment. *Journal of Environmental Quality* 46:1519-1525. doi: https://doi.org/10.2134/jeq2017.05.0202

2. **Zinabu, Eskinder** (2011) Assessment of the impact of industrial effluents on the quality of irrigation water and changes in soil characteristics: The case of Kombolcha town. *Journal of Irrigation and Drainage* 60:644-653
   doi: https://doi.org/10.1002/ird.609.

3. **Zinabu Eskinder**, Kelderman P, van der Kwast J, Kenneth I. Preventing sustainable development: policy and capacity gaps for monitoring metals in riverine water and sediments within an industrialising catchment in Ethiopia. *In submission*

Conference/workshop presentation

Eskinder Zinabu (2016) Catchment scale modelling of point and nonpoint sources nitrogen and phosphorus pollutants: applicable for data scarce sub-Saharan countries, October 3-4, Delft, The Netherlands.

Eskinder Zinabu (2015) Effects of point and non-point pollution sources on water quality in a River basin in Ethiopia: Result and Analyses of 2013 study works, Jan 8, 2015, Delft, The Netherlands.

Eskinder Zinabu (2014) Technological Solution for Waste Contamination and Water Pollution, the case of Kombolcha city, Ethiopia; International seminar at Galilee International Management Institution, June 14 - 28/2014, Galilee, Israel.

Eskinder Zinabu (2012) Modelling Industrial Effluents for Optimized Water Quality in Data-Poor Sub-Saharan Countries: the Case of Kombolcha, Ethiopia. 2012. Delft, The Netherlands.

Eskinder Zinabu (2012) Modelling Industrial Effluents for Optimized Water Quality in Data-Poor Sub-Saharan Countries: the Case of Kombolcha city, EthiopiaMarch.2012, Apeldoorn, The Netherlands.

Eskinder Zinabu (2010) Assessment of the impact of industrial effluents on the quality of irrigation water and changes on soil Characteristics (a case of Kombolcha town). Fourteenth International Water Technology Conference, IWTC 14 2010, Cairo, Egypt.

Netherlands Research School for the
Socio-Economic and Natural Sciences of the Environment

D I P L O M A

For specialised PhD training

The Netherlands Research School for the
Socio-Economic and Natural Sciences of the Environment
(SENSE) declares that

Eskinder Zinabu Belachew

born on 20 June 1974 in DebreBirhan, Ethiopia

has successfully fulfilled all requirements of the
Educational Programme of SENSE.

Delft, 26 March 2019

the Chairman of the SENSE board   the SENSE Director of Education

Prof.dr. Martin Wassen   Dr. Ad van Dommelen

The SENSE Research School has been accredited by the Royal Netherlands Academy of Arts and Sciences (KNAW)

K O N I N K L I J K E   N E D E R L A N D S E
A K A D E M I E   V A N   W E T E N S C H A P P E N

The SENSE Research School declares that Mr Eskinder Belachew has successfully fulfilled all requirements of the Educational PhD Programme of SENSE with a work load of 37 EC, including the following activities:

SENSE PhD Courses

- o Environmental research in context (2012)
- o Research in context activity: 'Co-organising seminar on catchment-scale surface water quality assessment to estimate industrial effluent loading into streams: the case of Kombolcha City', Ethiopia (2015)

Other PhD and Advanced MSc Courses

- o Data acquisition, preprocessing and modelling using the PCRaster Python framework, UNESCO-IHE, Delft (2013)
- o Water quality assessment, UNESCO-IHE, Delft (2013)

External training at a foreign research institute

- o Environmental Management, Galilee International Management Institution, Israel (2014)

Management and Didactic Skills Training

- o Assisting in the MSc course 'Water quality assessment', 2014-2016
- o Supervising two MSc students with thesis entitled 'Environmental impact of chromium and zinc from tannery and steel industries effluent in Leyole River, Ethiopia'(2013) and 'Effects of heavy metals from industries in river water and sediments for Kombolcha, Ethiopia' (2014)

Oral Presentations

- o *Modelling Industrial Effluents for Optimized Water Quality in Data-Poor Sub-Saharan Countries: the Case of Kombolcha City.* PhD seminar UNESCO-IHE, 1-2 October 2012, Delft, The Netherlands
- o *Technological Solution for Waste Contamination and Water Pollution, the case of Kombolcha city, Ethiopia.* International seminar Environmental Management, Galilee International Management Institution, 14-28 June 2014, Galilee, Israel
- o *Effects of point and non-point pollution sources on water quality in a River basin in Ethiopia.* Scientific meeting Aquatic ecosystems, UNESCO-IHE, 8 January 2015, Delft, The Netherlands

SENSE Coordinator PhD Education

Dr. P.J. Vermeulen